At Sylvan, we believe that a lifelong love of learning begins at an early age, and we are glad you have chosen our resources to help your children experience the joy of mathematics as they build critical reasoning skills. We know that the time you spend with your children reinforcing the lessons learned in school will contribute to their love of learning.

Success in math requires more than just memorizing basic facts and algorithms; it also requires children to make sense of size, shape, and numbers as they appear in the world. Children who can connect their understanding of math to the world around them will be ready for the challenges of mathematics as they advance to more complex topics.

We use a research-based, step-by-step process in teaching math at Sylvan that includes thought-provoking math problems and activities. As students increase their success as problem solvers, they become more confident. With increasing confidence, students build even more success. The design of the Sylvan workbooks will help you to help your children build the skills and confidence that will contribute to success in school.

Included with your purchase of this workbook is a coupon for a discount at a participating Sylvan center. We hope you will use this coupon to further your children's academic journeys. Let us partner with you to support the development of confident, well-prepared, independent learners.

The Sylvan Team

Sylvan Learning Center
Unleash your child's potential here

No matter how big or small the academic challenge, every child has the ability to learn. But sometimes children need help making it happen. Sylvan believes every child has the potential to do great things. And, we know better than anyone else how to tap into that academic potential so that a child's future really is full of possibilities. Sylvan Learning Center is the place where your child can build and master the learning skills needed to succeed and unlock the potential you know is there.

The proven, personalized approach of our in-center programs delivers unparalleled results that other supplemental education services simply can't match. Your child's achievements will be seen not only in test scores and report cards but outside the classroom as well. And when your child starts achieving his or her full potential, everyone will know it. You will see a new level of confidence come through in all of your child's activities and interactions.

How can Sylvan's personalized in-center approach help your child unleash the potential you know is there?

- Starting with our exclusive Sylvan Skills Assessment®, we pinpoint your child's exact academic needs.
- Then we develop a customized learning plan designed to meet your child's academic goals.
- Through our method of skill mastery, your child will not only learn and master every skill in a personalized plan, but he or she will be truly motivated and inspired to achieve.

To get started, included with this Sylvan product purchase is $10 off our exclusive Sylvan Skills Assessment®. Simply use this coupon and contact your local Sylvan Learning Center to set up your appointment.

To learn more about Sylvan and our innovative in-center programs, call 1-800-EDUCATE or visit www.SylvanLearning.com. ***With over 900 locations in North America, there is a Sylvan Learning Center near you!***

3rd Grade Geometry Success

Published in the United States by Random House, Inc., New York, and in Canada by Random House of Canada Limited, Toronto.

www.tutoring.sylvanlearning.com

Created by Smarterville Productions LLC
Producer & Editorial Direction: The Linguistic Edge
Producer: TJ Trochlil McGreevy
Writer: Kelly Woodard Parker
Cover and Interior Illustrations: Tim Goldman, Shawn Finley, and Duendes del Sur
Layout and Art Direction: SunDried Penguin

First Edition

ISBN: 978-0-307-47928-0
ISSN: 2156-5996

This book is available at special discounts for bulk purchases for sales promotions or premiums. For more information, write to Special Markets/Premium Sales, 1745 Broadway, MD 6-2, New York, New York 10019 or e-mail specialmarkets@randomhouse.com.

PRINTED IN CHINA

10 9 8 7 6 5 4 3 2 1

Contents

Plane Shapes

Solid Shapes

Patterns & Sets

Symmetry

Flips, Slides & Turns

Perimeter & Area

Contents

Shape Puzzles

What's My Name?

Plane shapes are two-dimensional shapes. Their names tell the number of sides they have. A **square** is a special type of rectangle.

A **triangle** has 3 sides.

A **rectangle** has 4 sides.

A **square** has 4 equal sides.

A **hexagon** has 6 sides.

An **octagon** has 8 sides.

WRITE the name of each shape.

1. ____________________

2. ____________________

3. ____________________

4. ____________________

5. ____________________

Matched Set

DRAW one or more lines from each word to each shape that is a match.

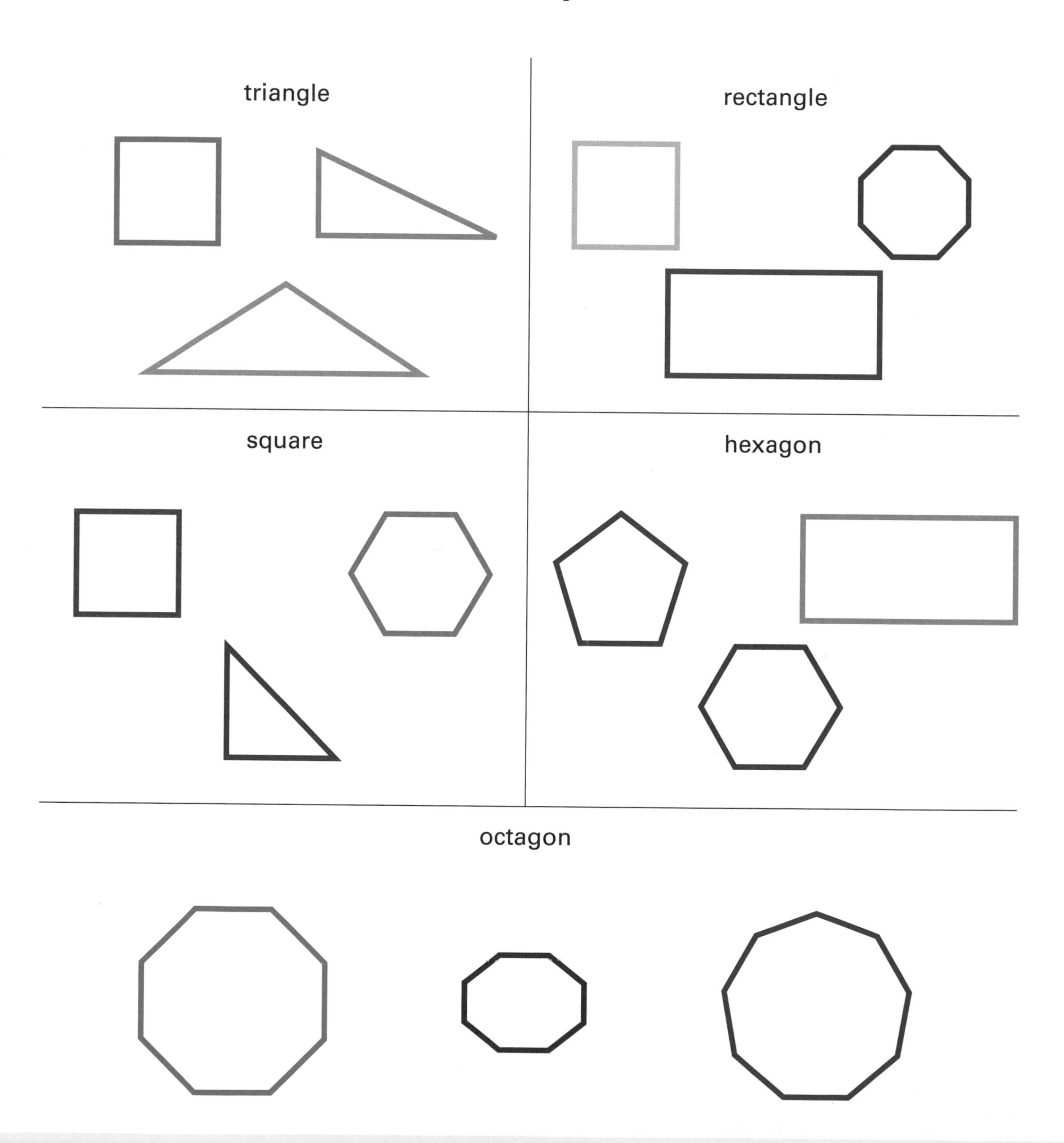

Odd One Out

CROSS OUT the shape in each row that is **not** the same type of shape as the others.

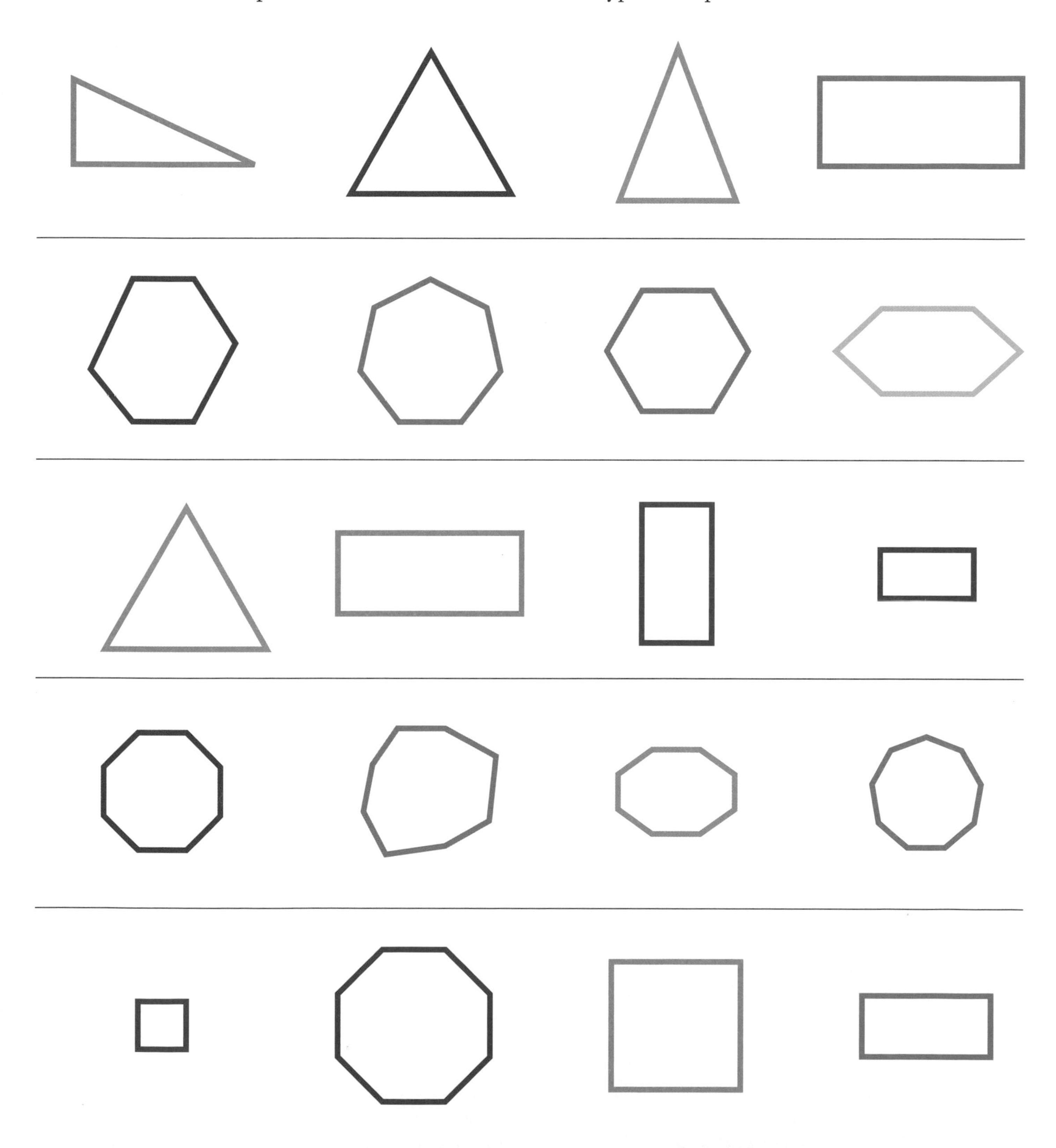

Hidden Words

WRITE the name of each shape. WRITE the numbered letters in numerical order to find the answer to the riddle.

1.

2.

3.

4.

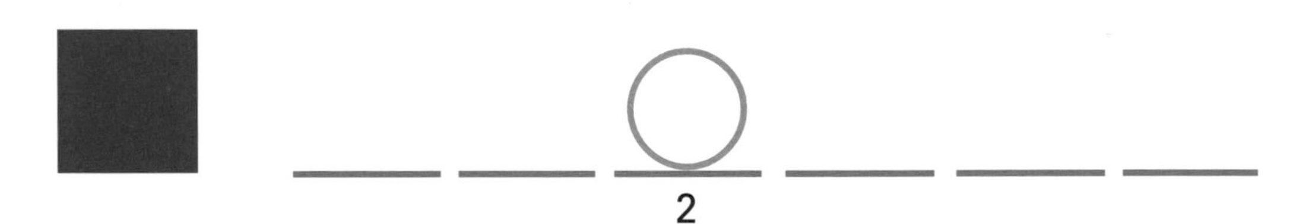

5.

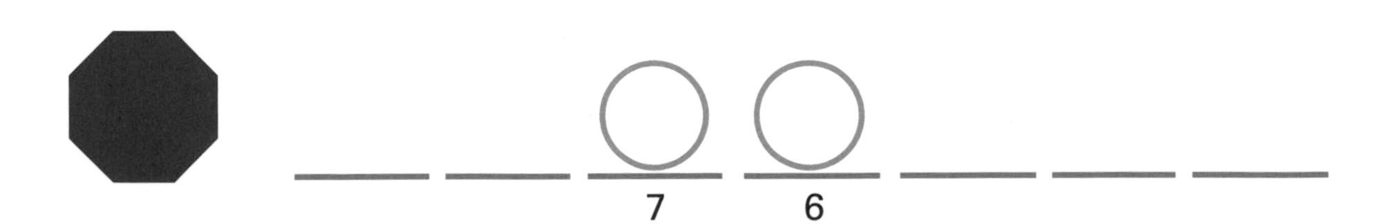

What can we all hold without using our hands or our arms?

___ ___ ___ B ___ ___ ___ ___ ___.

1 2 3 4 5 6 7 8

Write It

The point where two lines meet is the **vertex.** The line segment between the vertices is a **side**.

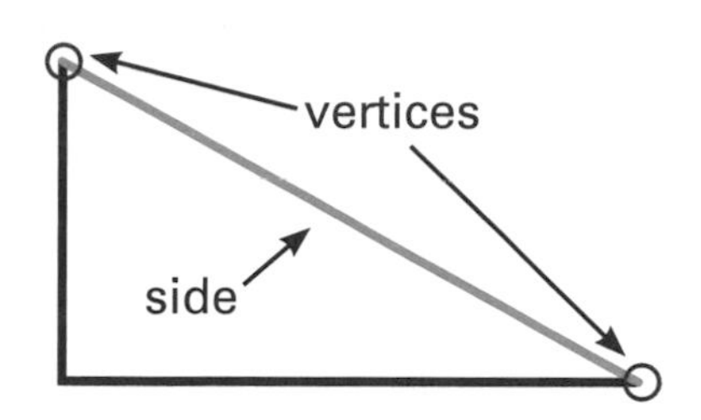

WRITE the name of each shape, CIRCLE each vertex, and PUT an X on each side.

1. ____________________

2. 

3. ____________________

4. 

5. ____________________

6. ____________________

Shape Up

DRAW or WRITE the missing information.

Shape	Number of Vertices	Number of Sides
	3	3
	6	6
	8	8

Odd One Out

CROSS OUT the description or shape in each row that is **not** the same type of shape as the others.

4 vertices
4 sides

3 vertices
3 sides

8 vertices
8 sides

7 vertices
7 sides

Circle the Same

CIRCLE all of the shapes and words that match each column's description.

3 Vertices 3 Sides	4 Vertices 4 Sides	6 Vertices 6 Sides	8 Vertices 8 Sides
	square	triangle	
			hexagon
octagon			

What's My Shape?

CIRCLE the shape that each person is describing.

I'm thinking of a shape that has six sides.

I'm thinking of a rectangle.

I'm thinking of a shape that has three vertices.

I'm thinking of an octagon.

Match Up

DRAW a line to match each shape to the same type of shape.

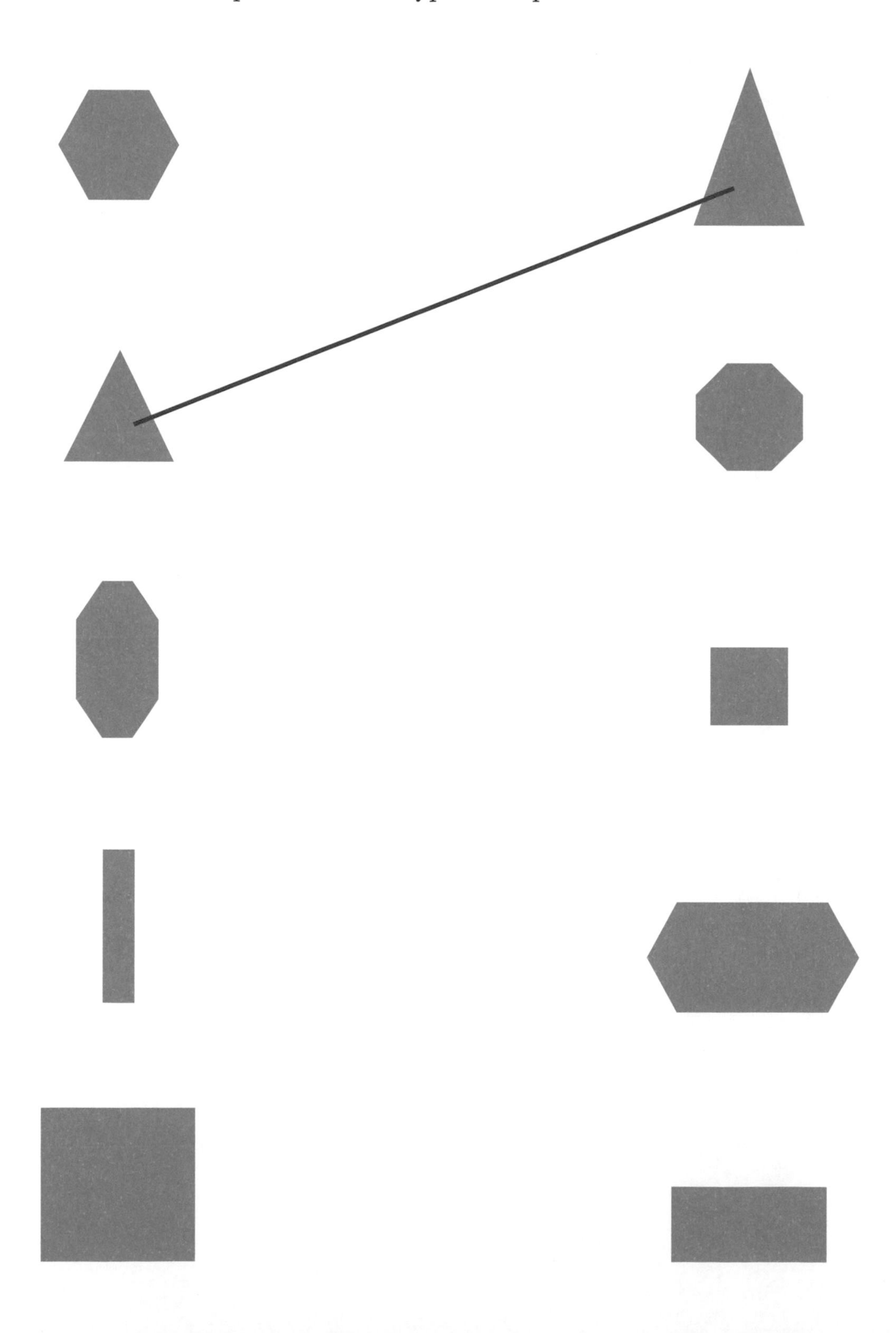

Pick a Shape

WRITE the letter of each shape that answers each question.

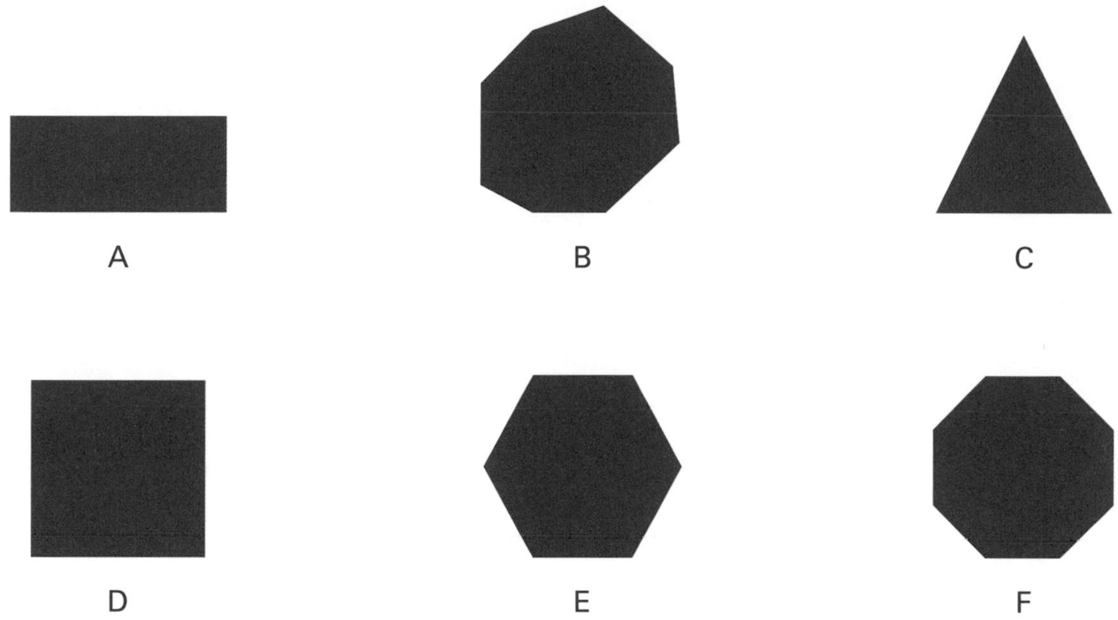

A B C

D E F

1. Which two shapes are octagons? ______
2. Which shape is a hexagon? ______
3. Which shape is a square? ______
4. Which shape has the fewest vertices? ______
5. Which two shapes are rectangles? ______
6. Which shape has the fewest sides? ______

Same or Different?

WRITE one similarity and one difference between the shapes in each row.

HINT: Use words such as *sides* and *vertices*.

Example:

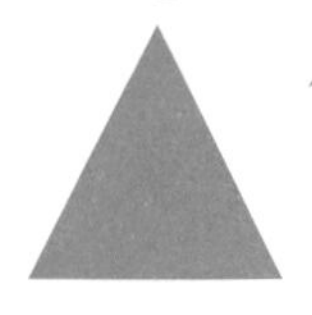
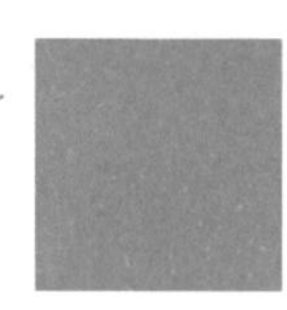

Similarity: The triangle and the square both have fewer than five vertices.

Difference: The triangle has three sides. The square has four sides.

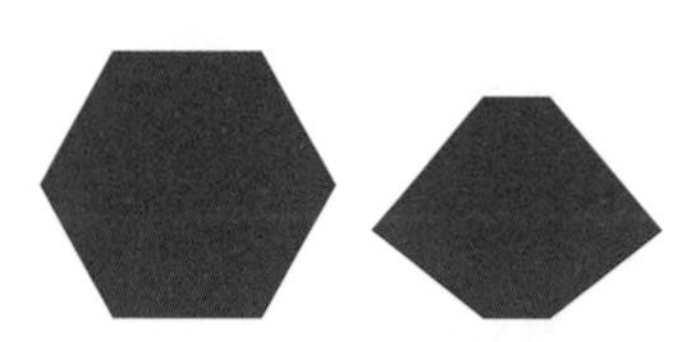

1. Similarity: ______________________________
____

Difference: ______________________________
____

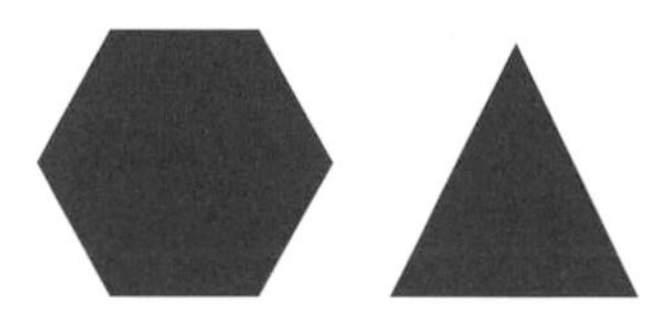

2. Similarity: ______________________________
____

Difference: ______________________________
____

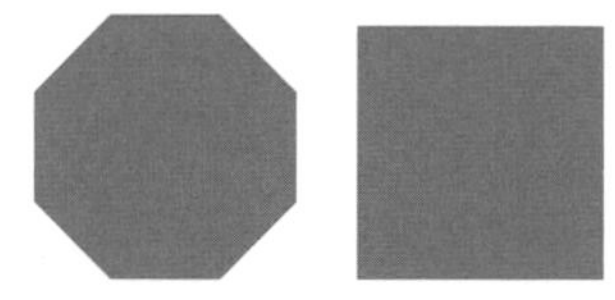

3. Similarity: ______________________________
____

Difference: ______________________________
____

4. Similarity: ______________________________
____

Difference: ______________________________
____

Dot Connector

CONNECT the dots to see how many different shapes you can create. Create at least one triangle, rectangle, square, hexagon, and octagon. Dots can be used in more than one shape, and shapes may overlap.

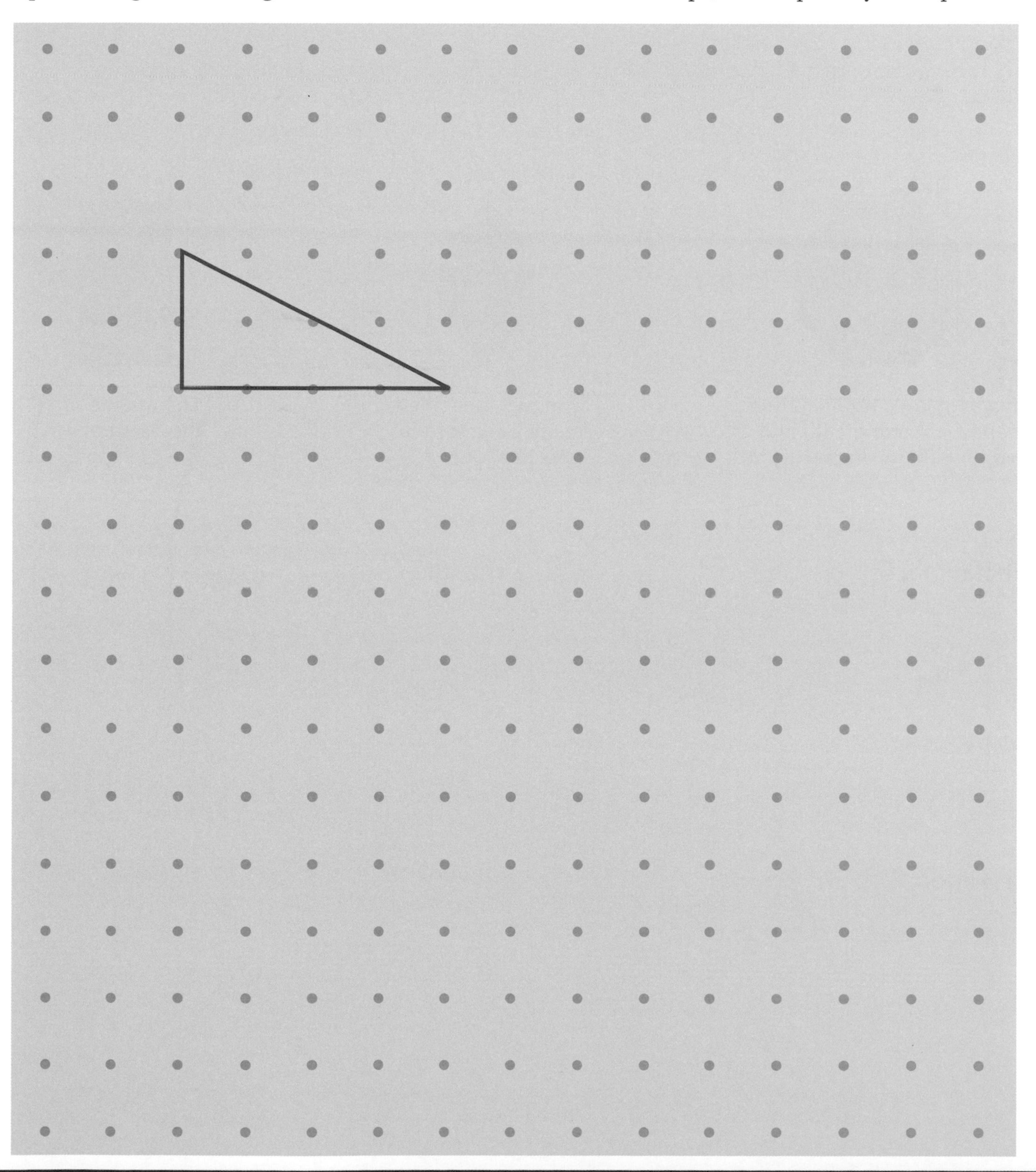

Shape Puzzler

CUT OUT the shapes on the following page. Then CUT each shape into smaller shapes, and REARRANGE the pieces to create new shapes. Each shape must be made from at least two parts. DRAW the new shape.

HINT: Remember that a hexagon can be any six-sided figure. The sides do not have to be the same length.

Example:

This octagon contains three shapes: a rectangle (L) and two four-sided shapes (K, M).

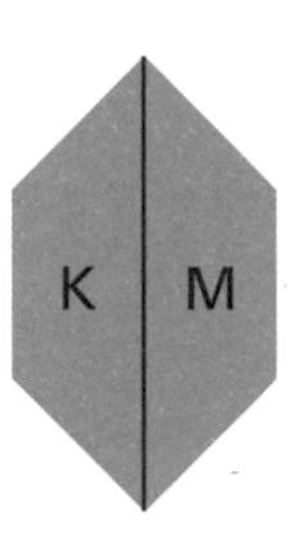

The two four-sided shapes can form a new hexagon.

The middle of the octagon, L, is a rectangle.

1. rectangle

2. triangle

3. hexagon

4

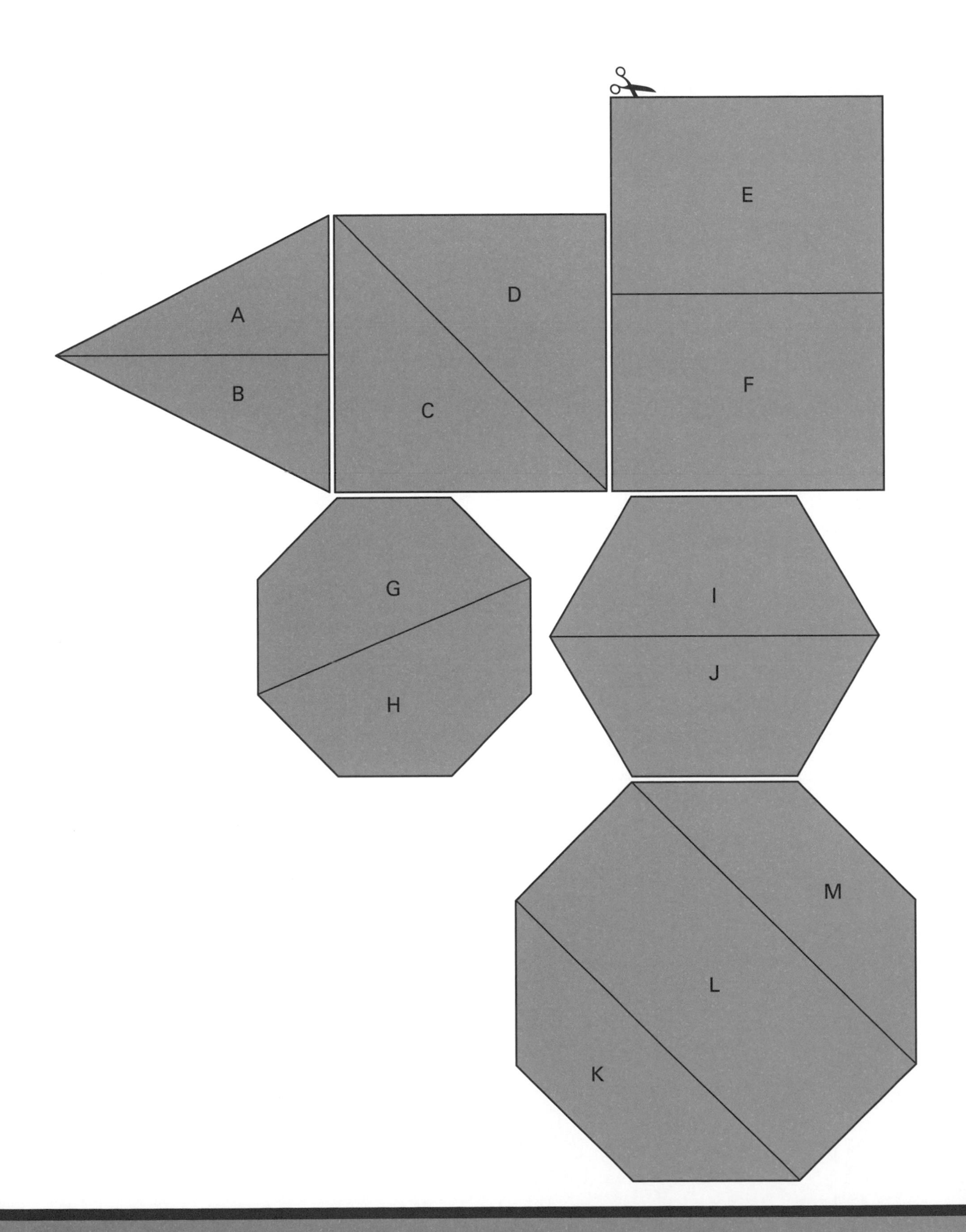
A
B
C
D
E
F
G
H
I
J
K
L
M

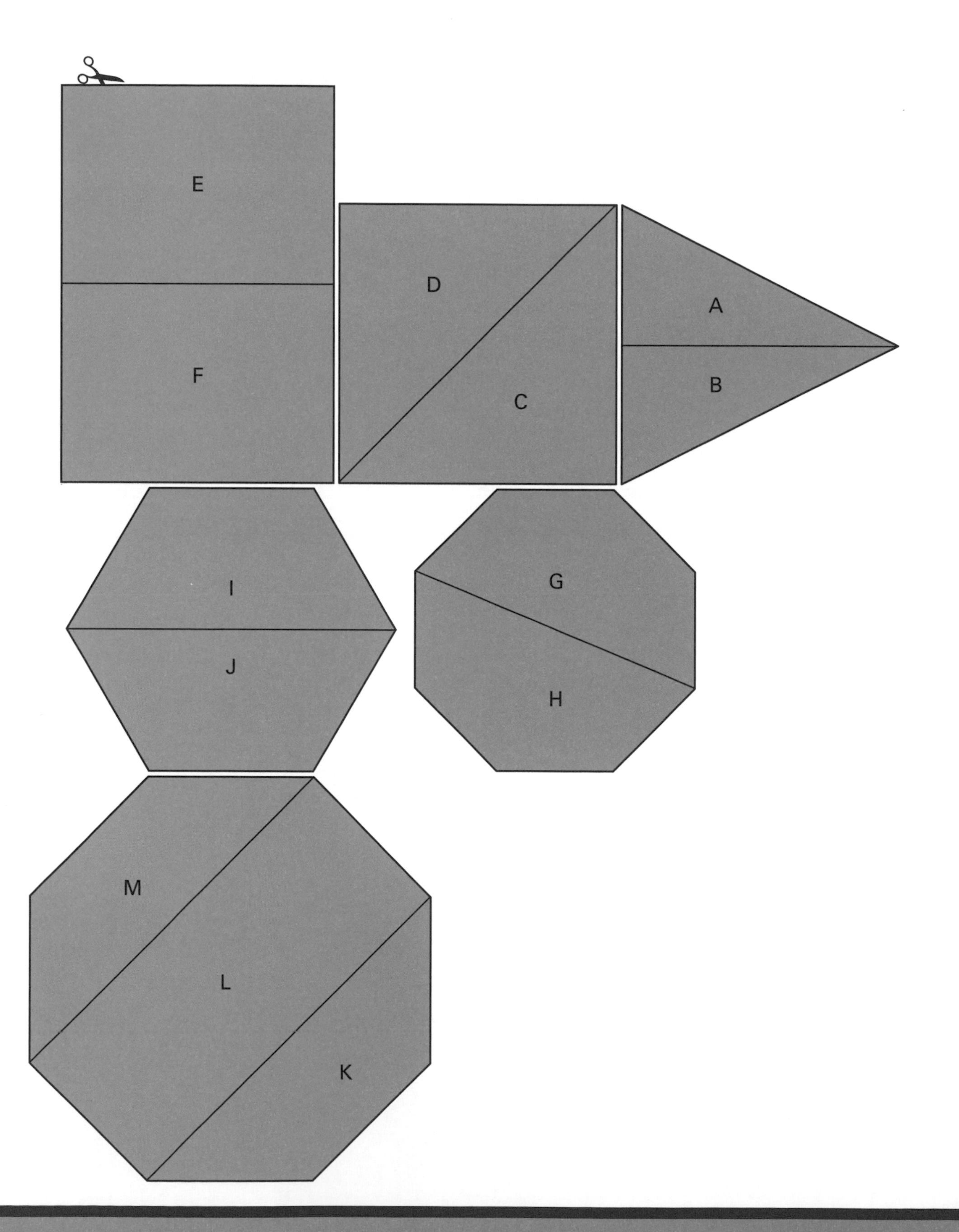
E
F
D
C
A
B
I
J
G
H
M
L
K

Pick a Shape

WRITE the letter of each shape that answers each question.

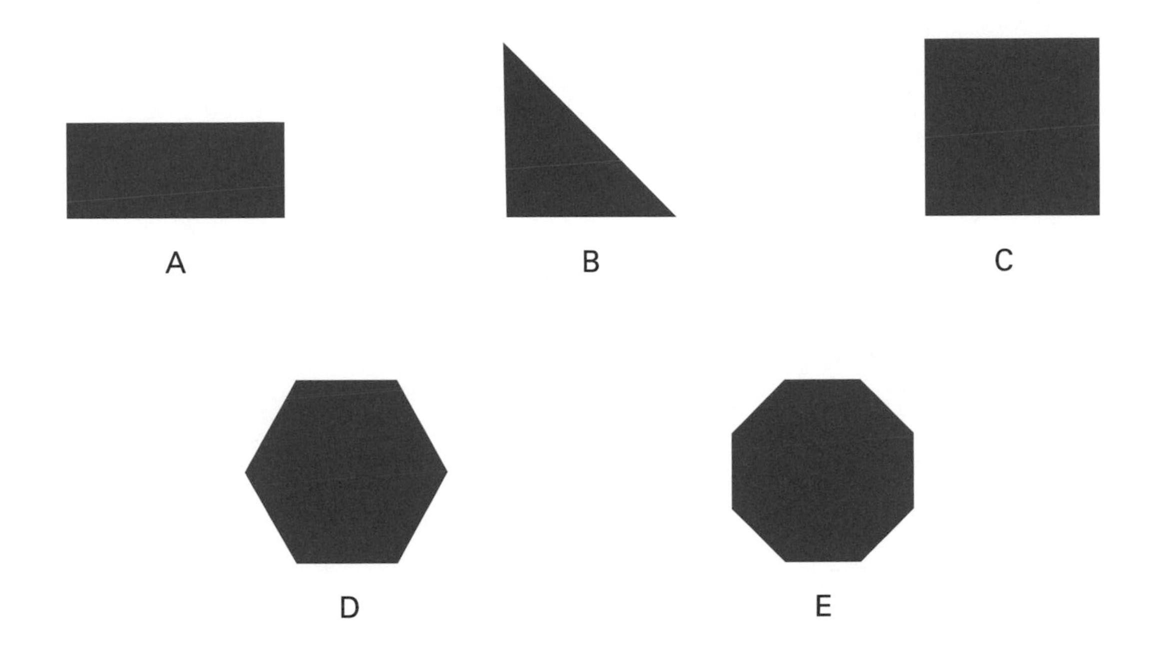

1. Which shape is an octagon? ______

2. Which shape is a square? ______

3. Which shapes have fewer sides than the hexagon? ______

4. Which two shapes have four sides and four vertices? ______

5. Which shape has three vertices? ______

6. Which shape has six sides? ______

Match Up

DRAW a line to match each shape to the same type of shape.

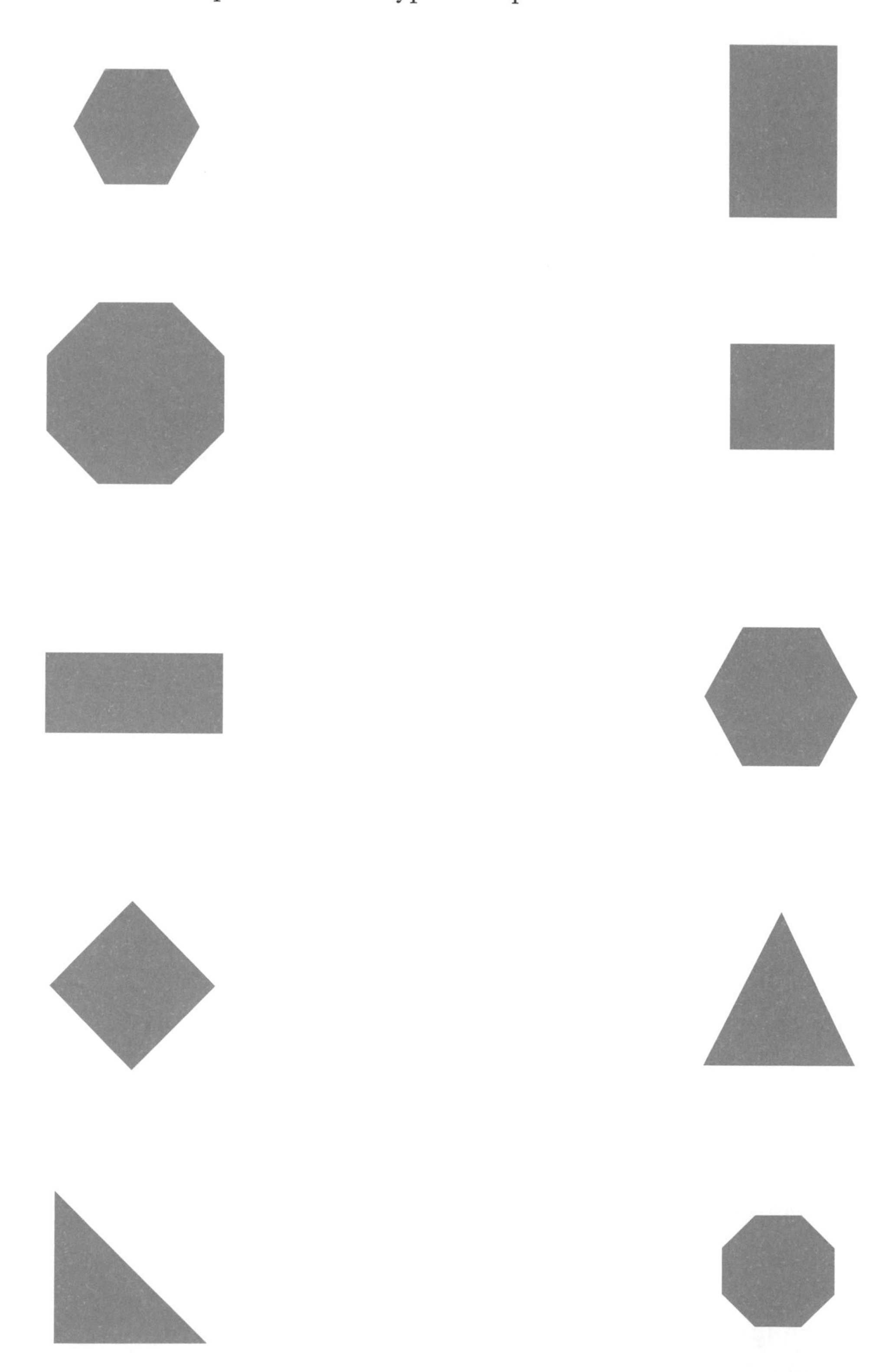

Shape Up

DRAW or WRITE the missing information.

Shape	Number of Vertices	Number of Sides
	6	6
	3	3
	8	8

Dot Connector

CONNECT dots to create a triangle, rectangle, square, hexagon, and octagon. Dots can be used in more than one shape, and shapes may overlap.

What's My Name?

A **solid shape** is a three-dimensional shape. A **cube** is a special rectangular prism. WRITE the name of each solid shape.

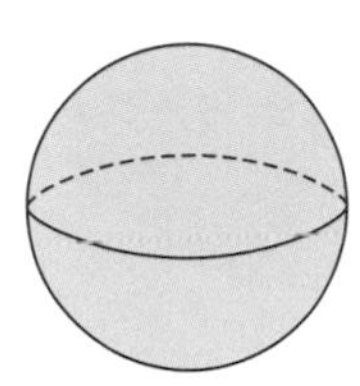
sphere

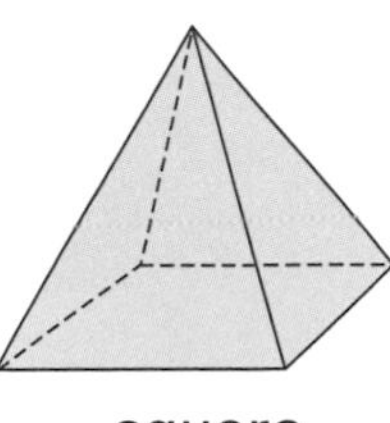
square pyramid

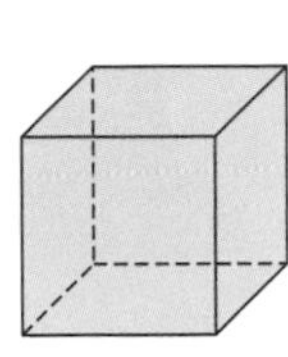
cube

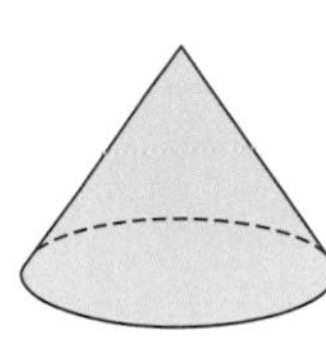
cone

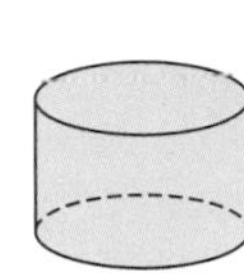
cylinder

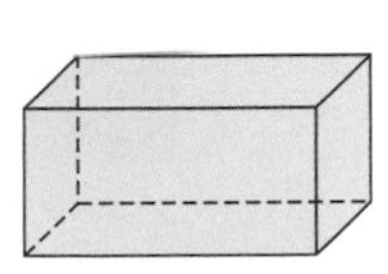
rectangular prism

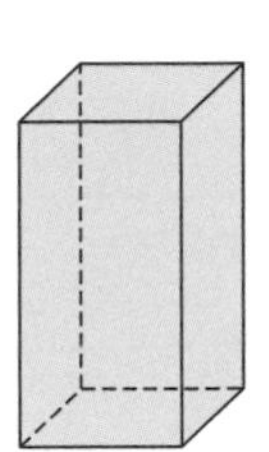

1. ______________________

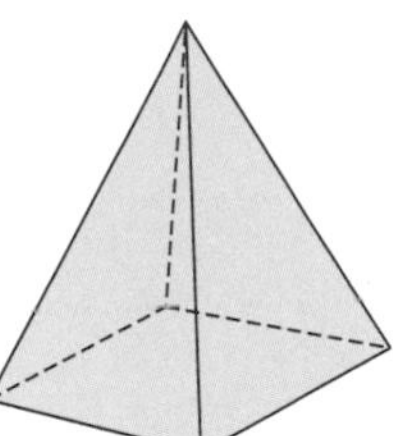
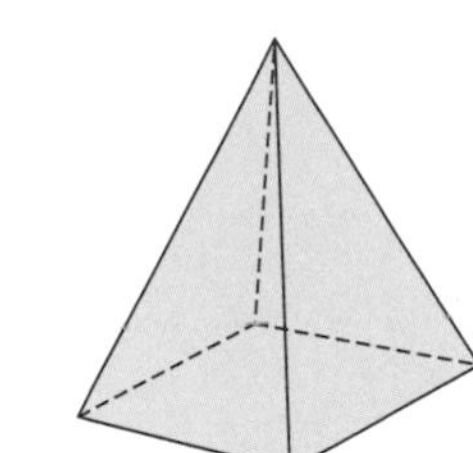

2. ______________________

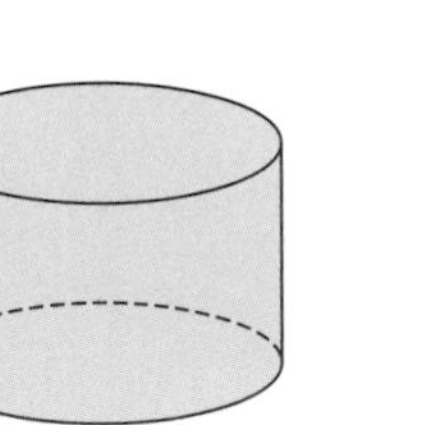

3. ______________________

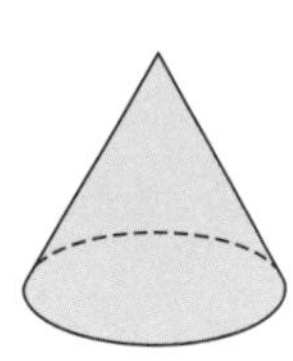

4. ______________________

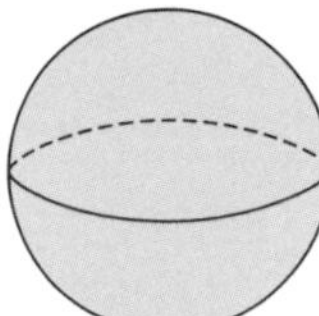
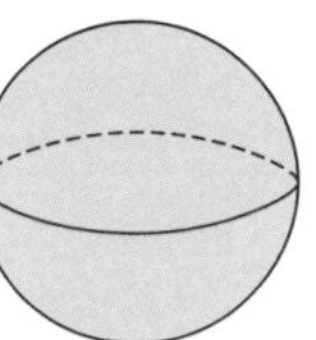

5. ______________________

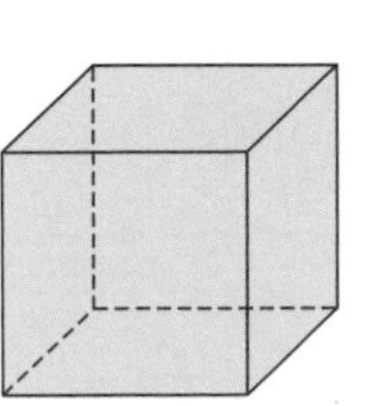

6 . ______________________

Matched Set

In each row, CIRCLE all of the pictures that match the name.

cube

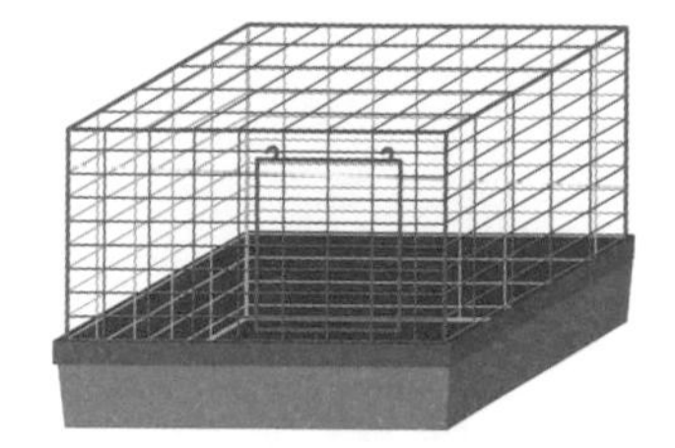

rectangular prism

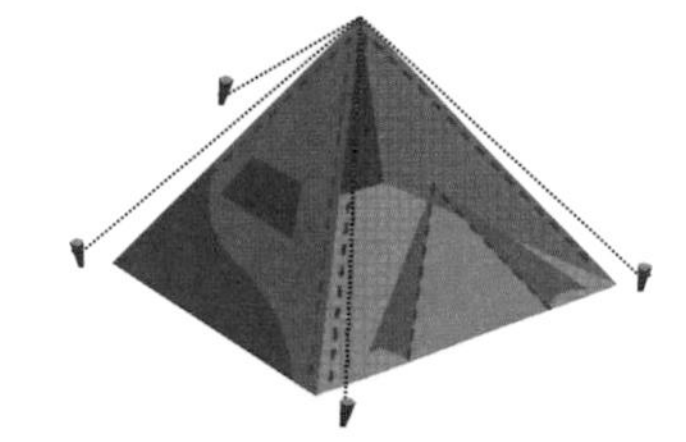

cone

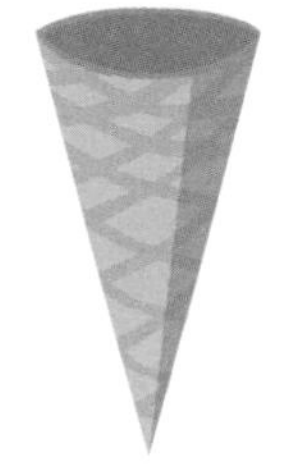

cylinder

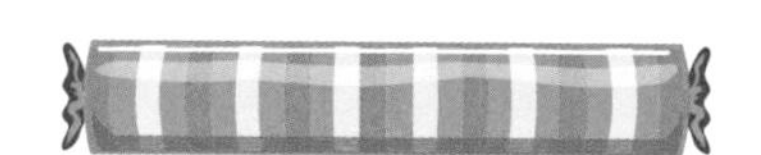

sphere

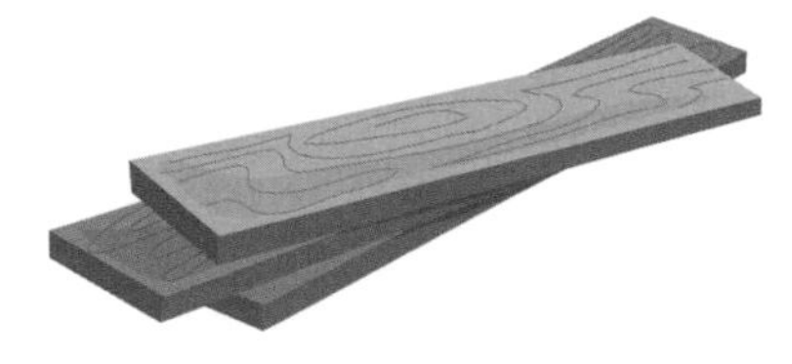

square pyramid

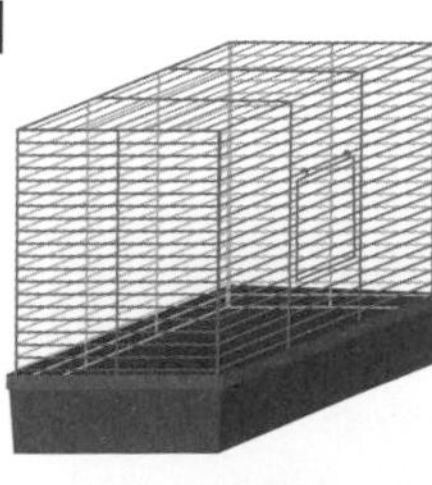
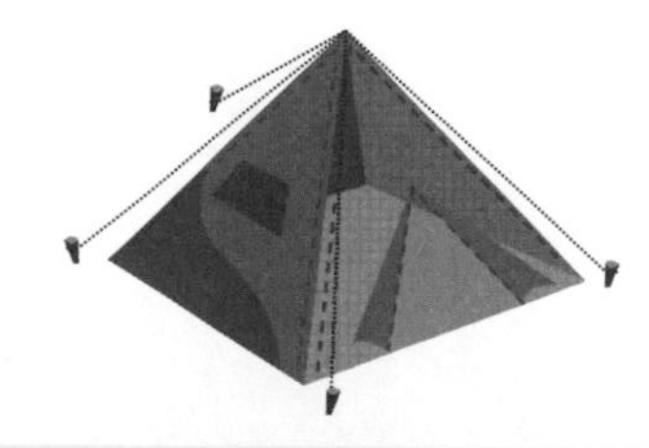

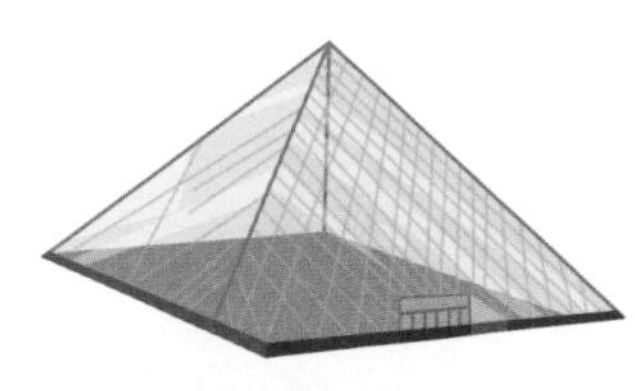

Party On

CIRCLE the solid shape that each party item most closely resembles.

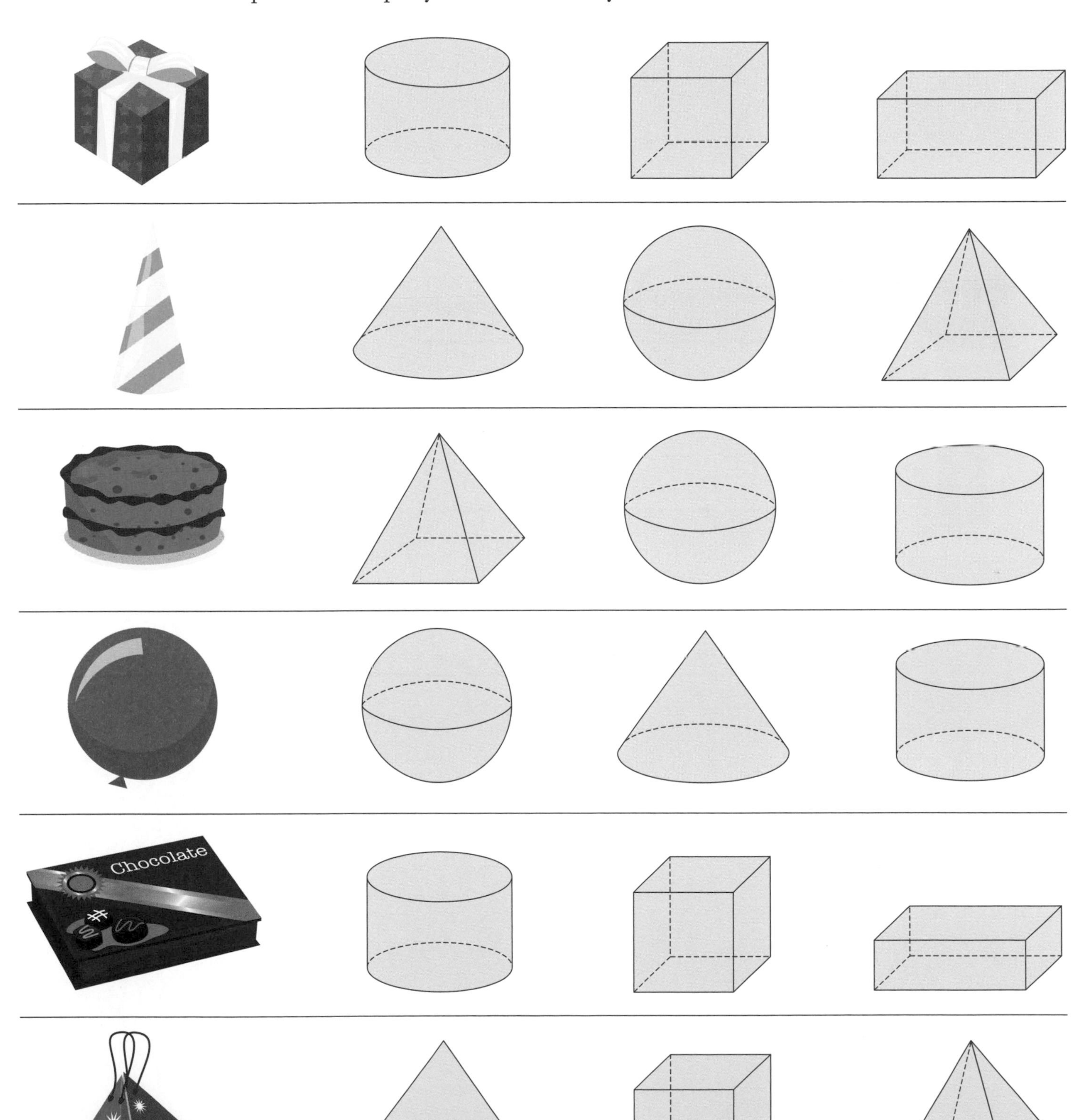

Hidden Words

WRITE the name of each solid shape. Then WRITE the numbered letters in numerical order to find the answer to the riddle.

1. 

___ ___ ___ ___ ___ ___ ___
9 ___ 1 4 ___ 11 ___

2. 

___ ___ ___ ___
___ 6 ___ ___

3. 

___ ___ ___ ___ ___ ___ ___ ___
___ ___ 7 ___ ___ ___ ___ 8

4. 

___ ___ ___ ___ ___ ___
12 ___ ___ 2 10 ___

5. 

___ ___ ___ ___
3 ___ 5 ___

A cube is a special

___ ___ ___ T ___ ___ G ___ ___ A ___
1 2 3 T 4 5 G 6 7 A 8

___ ___ ___ ___ M.
9 10 11 12 M.

Shape Up

Solid shapes have **edges**, which is where two faces meet. They have **vertices**, where three or more edges meet, and **faces**, which are the plane shapes formed by the edges.

Example:

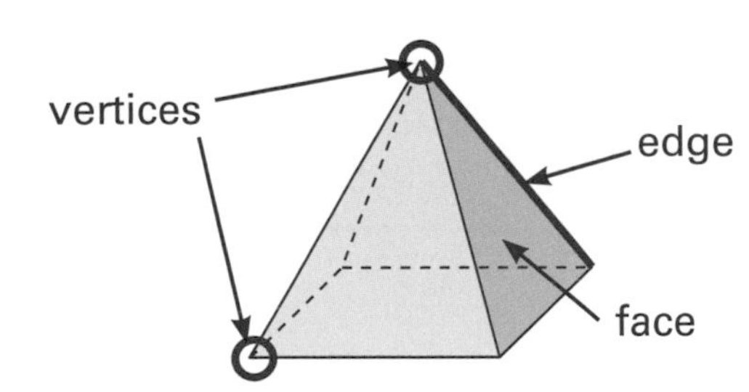

WRITE the number of edges, vertices, and faces for each solid shape.

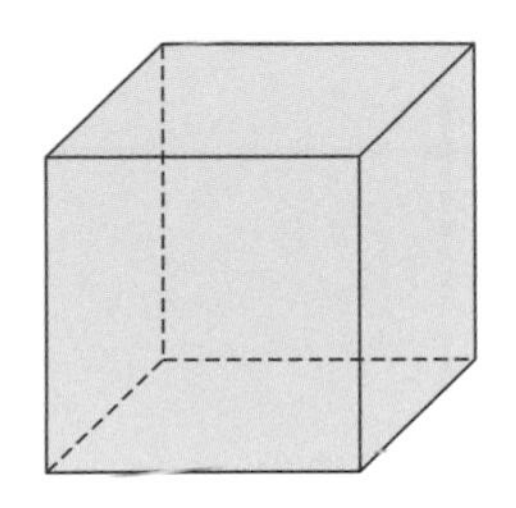

1. ______ edges ______ vertices ______ faces

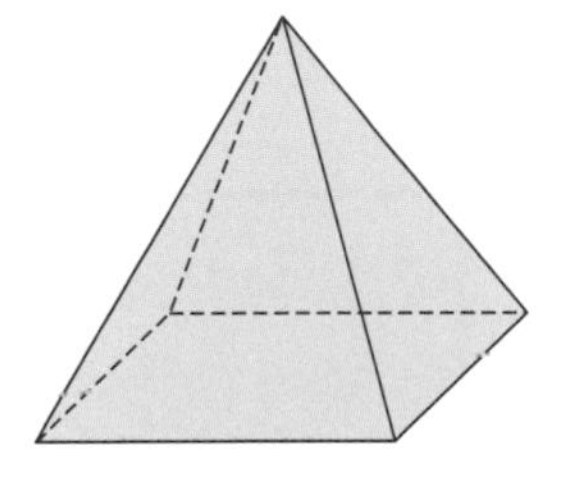

2. ______ edges ______ vertices ______ faces

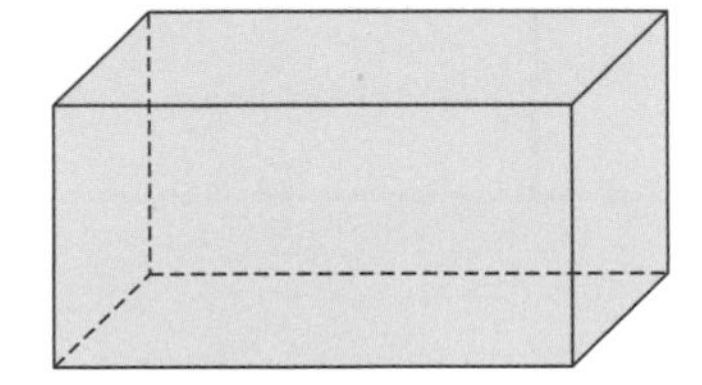

3. ______ edges ______ vertices ______ faces

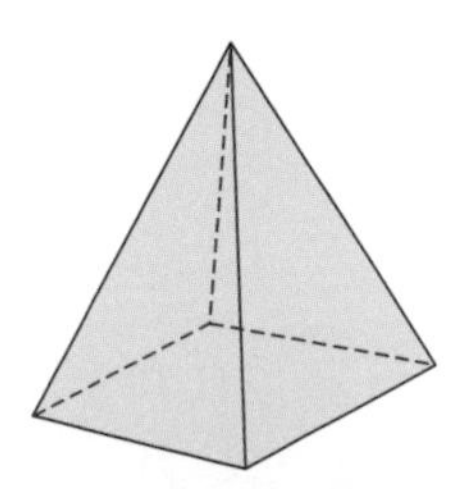

4. ______ edges ______ vertices ______ faces

Shape Up

WRITE the number of edges, vertices, and faces for each picture.

	Number of Edges	Number of Vertices	Number of Faces
1. SUGAR CUBES			
2.			
3. Loopy Fruits			

About Face

CIRCLE all of the shapes that are faces on the three-dimensional shape. If there is a face missing, draw the missing shape.

				Missing Face

Circle the Same

CIRCLE the three-dimensional shape that could be created by folding each drawing along the dotted lines.

Example:

square pyramid

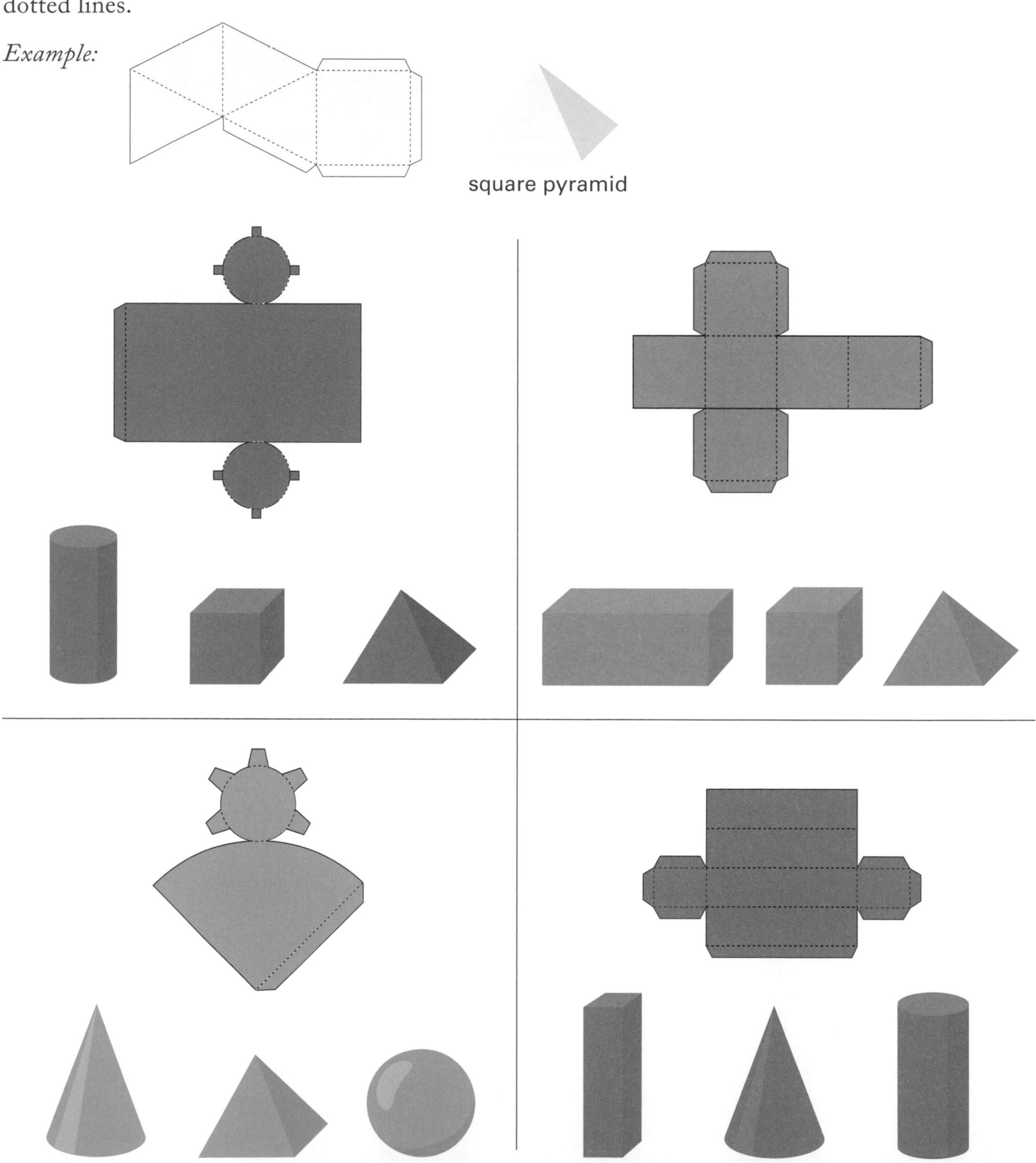

Pick a Shape

WRITE the letter of each solid shape that answers each question.

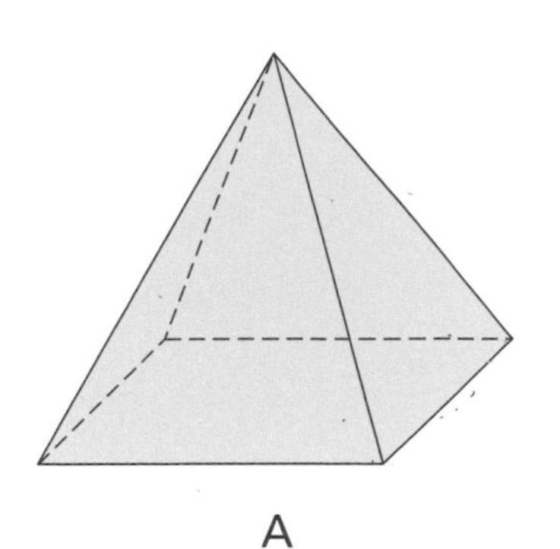
A

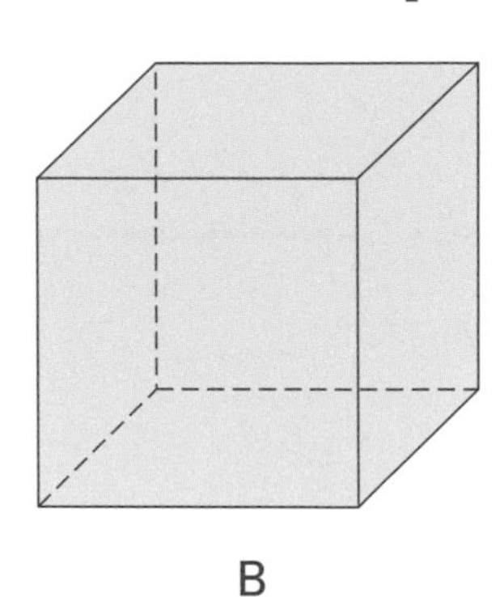
B

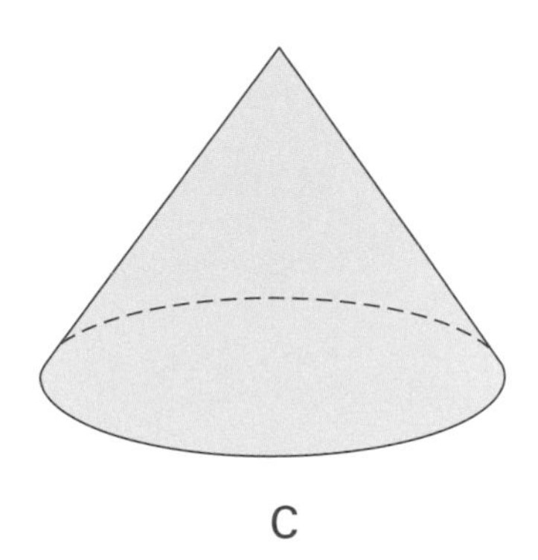
C

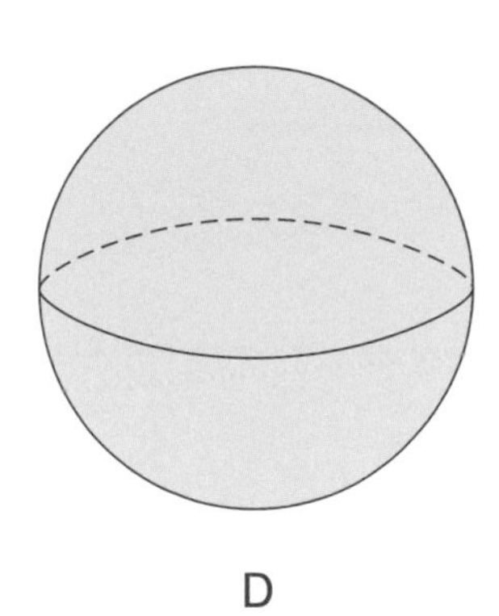
D

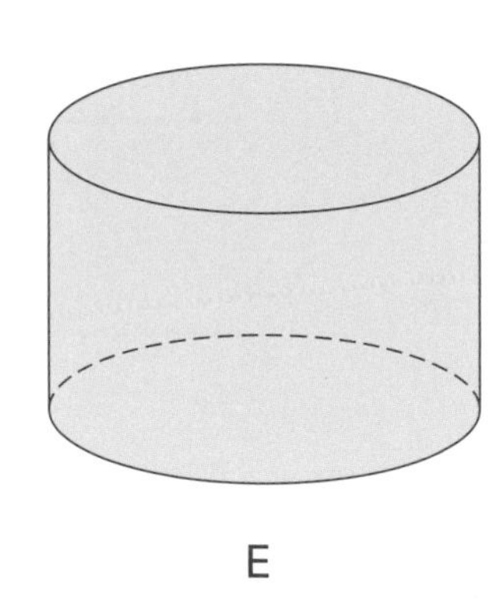
E

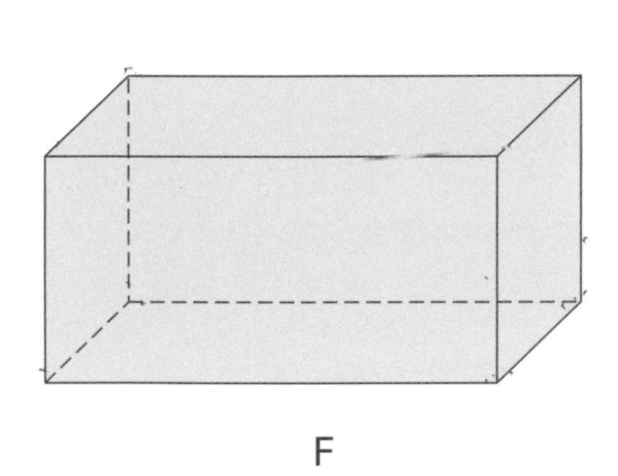
F

1. Which solid shape is a sphere? ____________
2. Which solid shapes have more than 5 faces? ____________
3. Which solid shape is a cone? ____________
4. Which solid shape has 5 vertices? ____________
5. Which solid shapes have 12 edges? ____________

Shape Builders

CUT OUT each shape on the opposite page. FOLD on the dotted lines, and GLUE the tabs to construct each solid shape. Then WRITE the answers.

1. What color is the cube? ______________________

2. How many vertices does each solid shape have:

   Blue: ____________ Yellow: ____________ Red: ____________

3. Which solid shape has the fewest faces? ______________________

4. What colors are the rectangular prisms? ______________________

5. Name one similarity between the blue and red solid shape.

   ____

6. Name one difference between the blue and yellow solid shape.

   ____

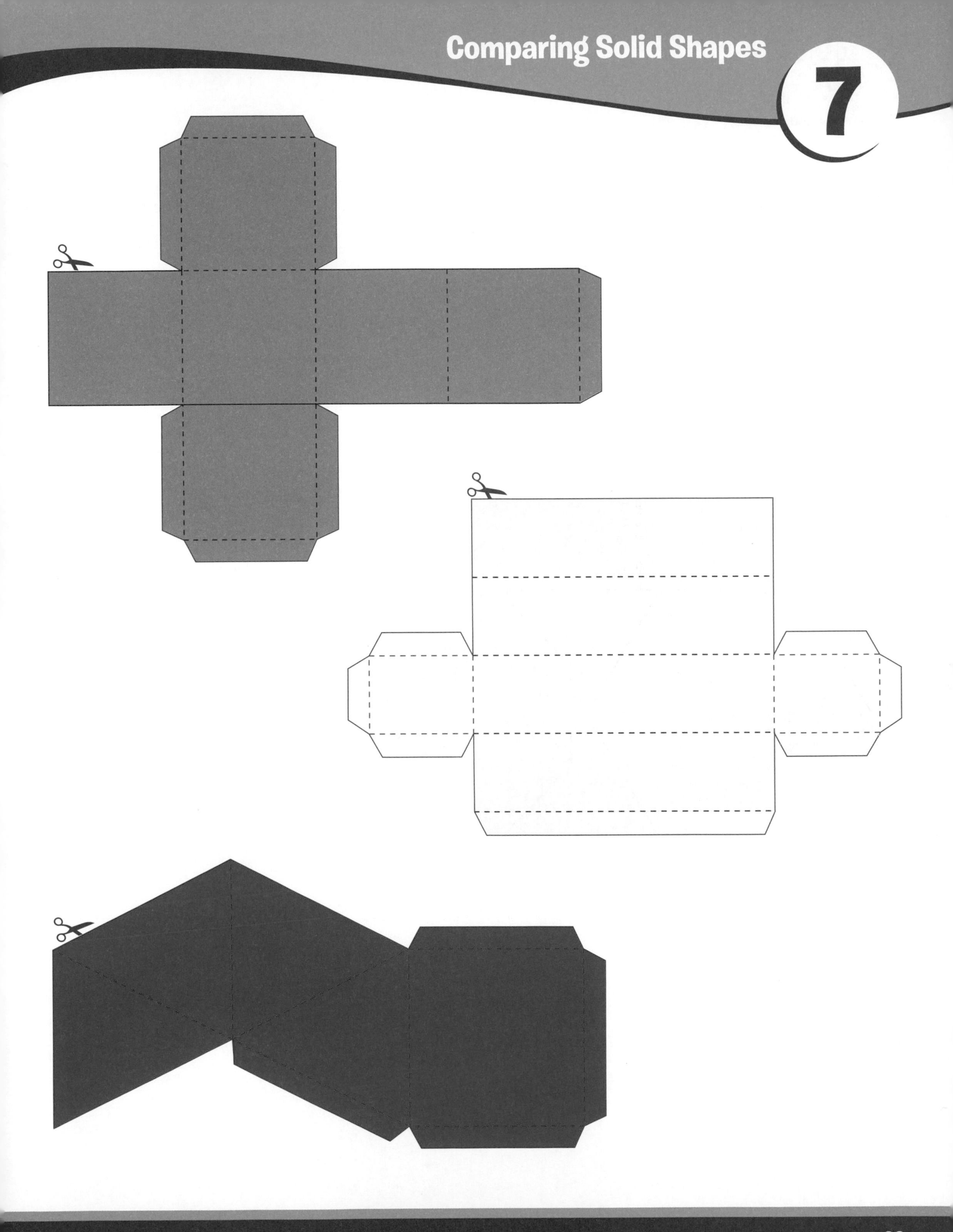

Match Up

DRAW a line to match each solid shape with its parts.

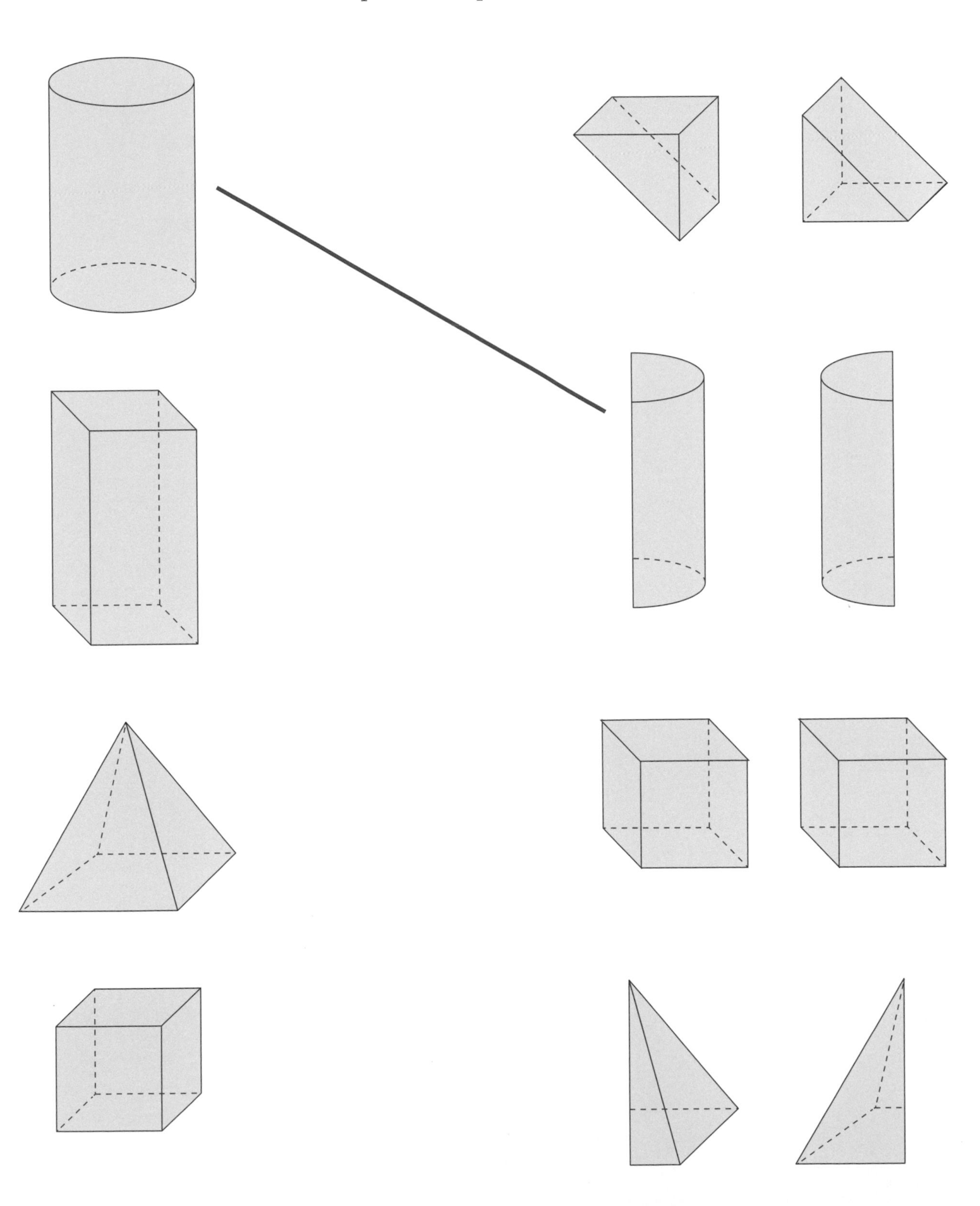

Shape Sums

WRITE the parts from the opposite page that create each solid shape.

Example: When parts W and Q are put together, the result is this rectangular prism.

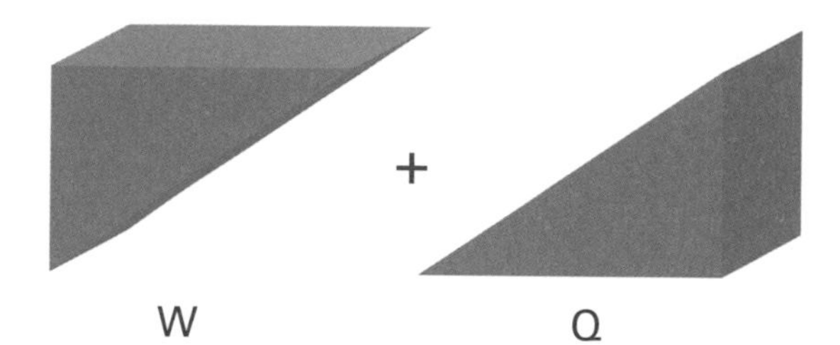

=

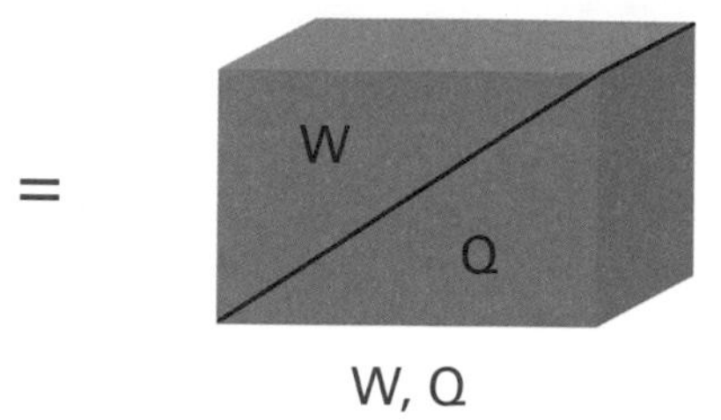

1. ____________

2. ____________

3. ____________

4. ____________

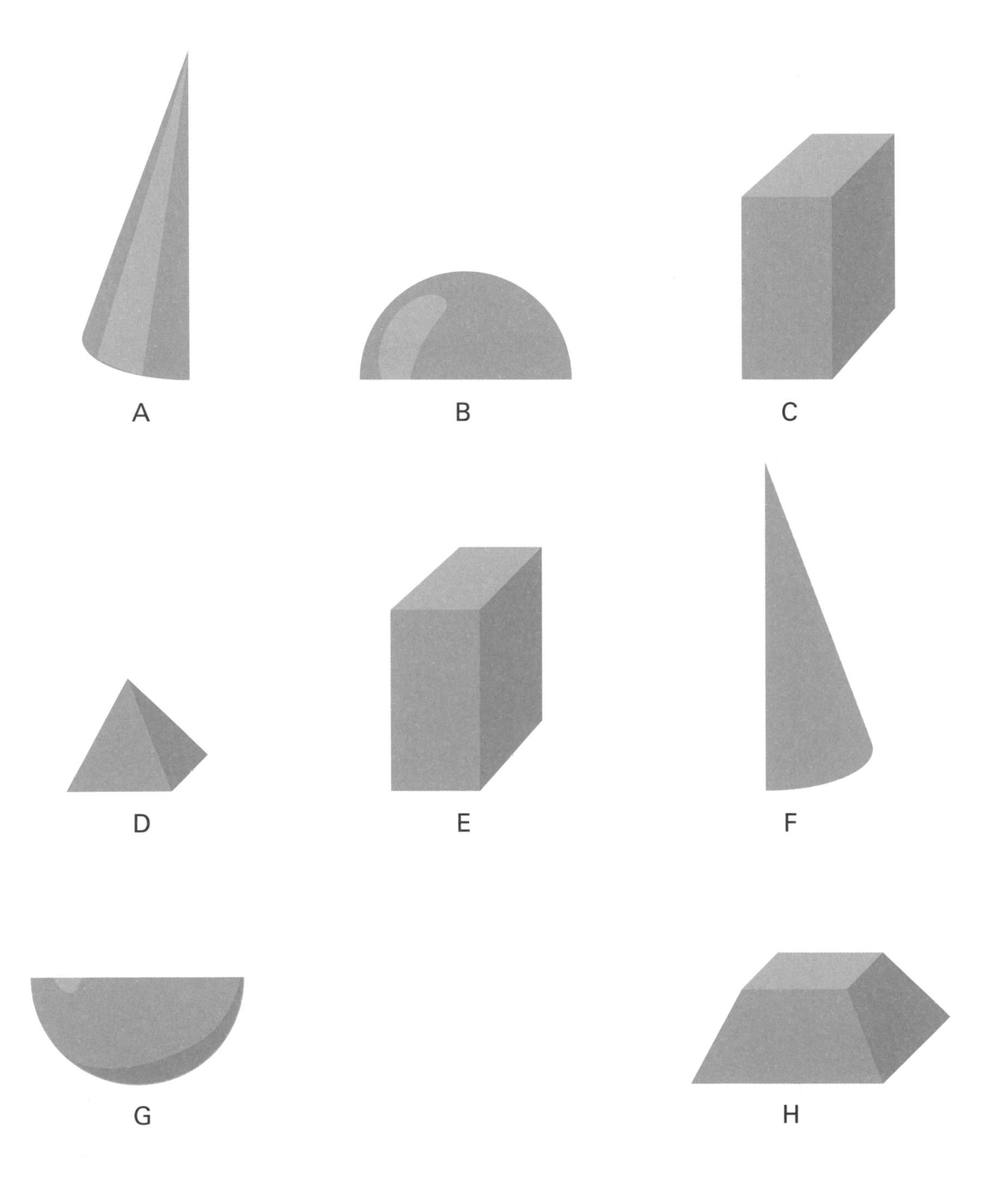
A
B
C
D
E
F
G
H

Shape Sums

WRITE the name of the solid shape or shapes needed to create the original picture.

1.  =  + ____________

2.  =  + ____________

3.  =  + ____________

4.  = + ____________

Match Up

DRAW a line to match each picture to the solid shape it most closely resembles.

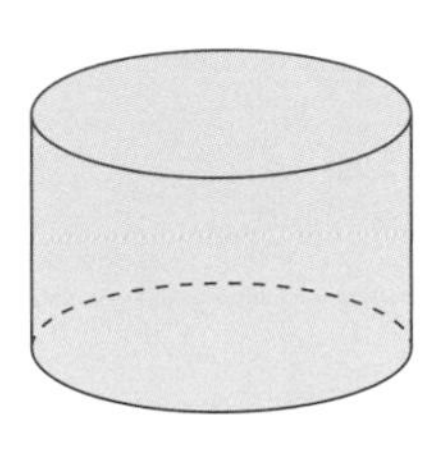

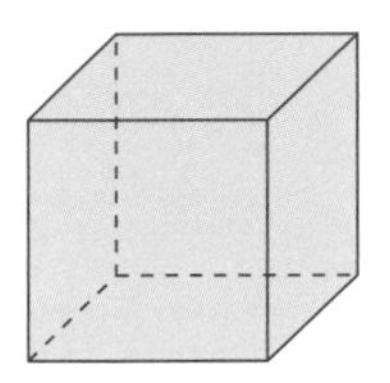

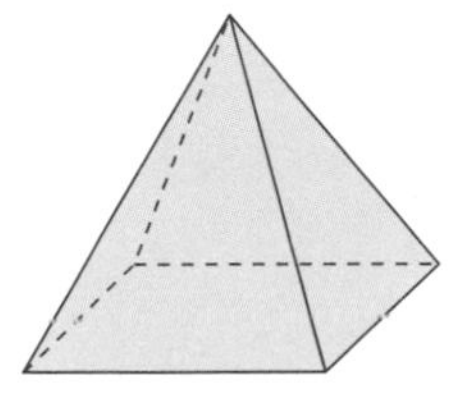

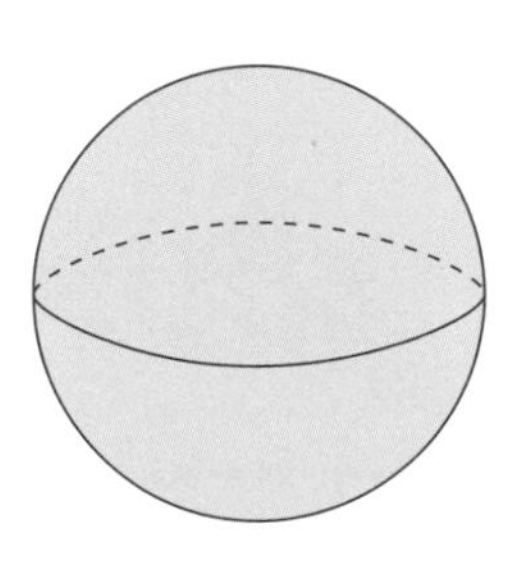

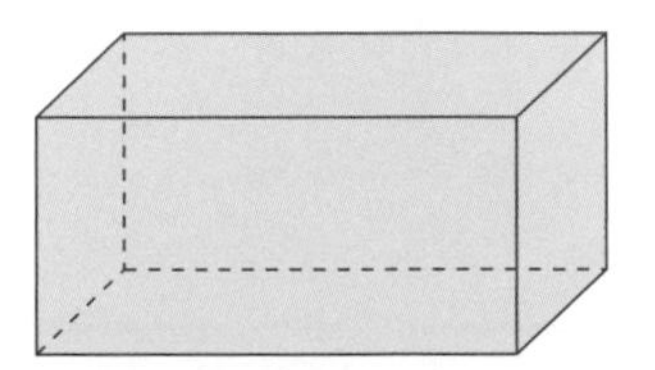

Shape Up

WRITE the name of the solid shape of each picture and the number of its edges, vertices, and faces.

	Name	Number of Edges	Number of Vertices	Number of Faces
 Congratulations!				
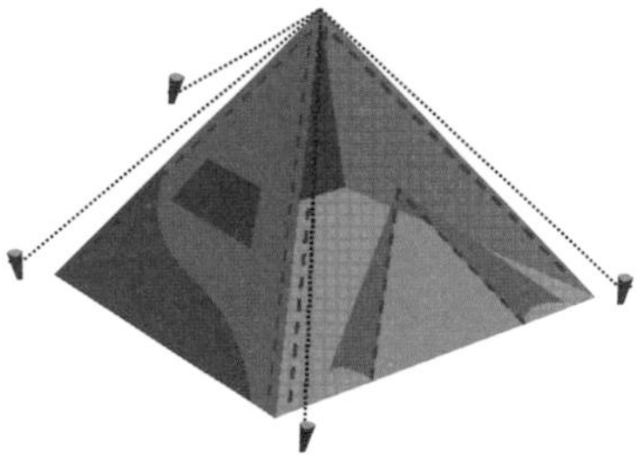				
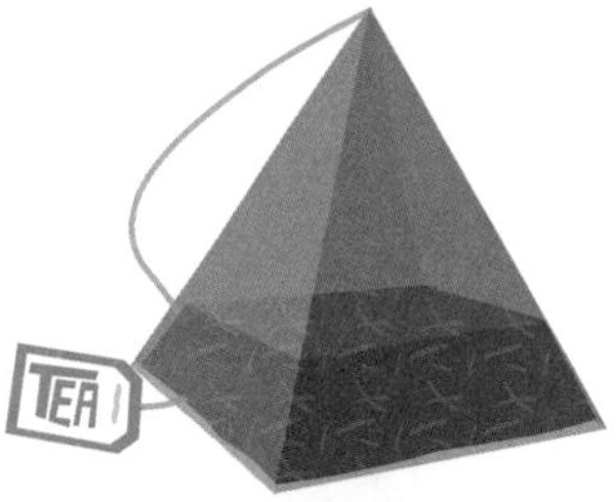 TEA				

About Face

DRAW a line or lines from each three-dimensional shape to each face that makes up the shape.

HINT: Some shapes use more than one face. Some faces are used on more than one shape.

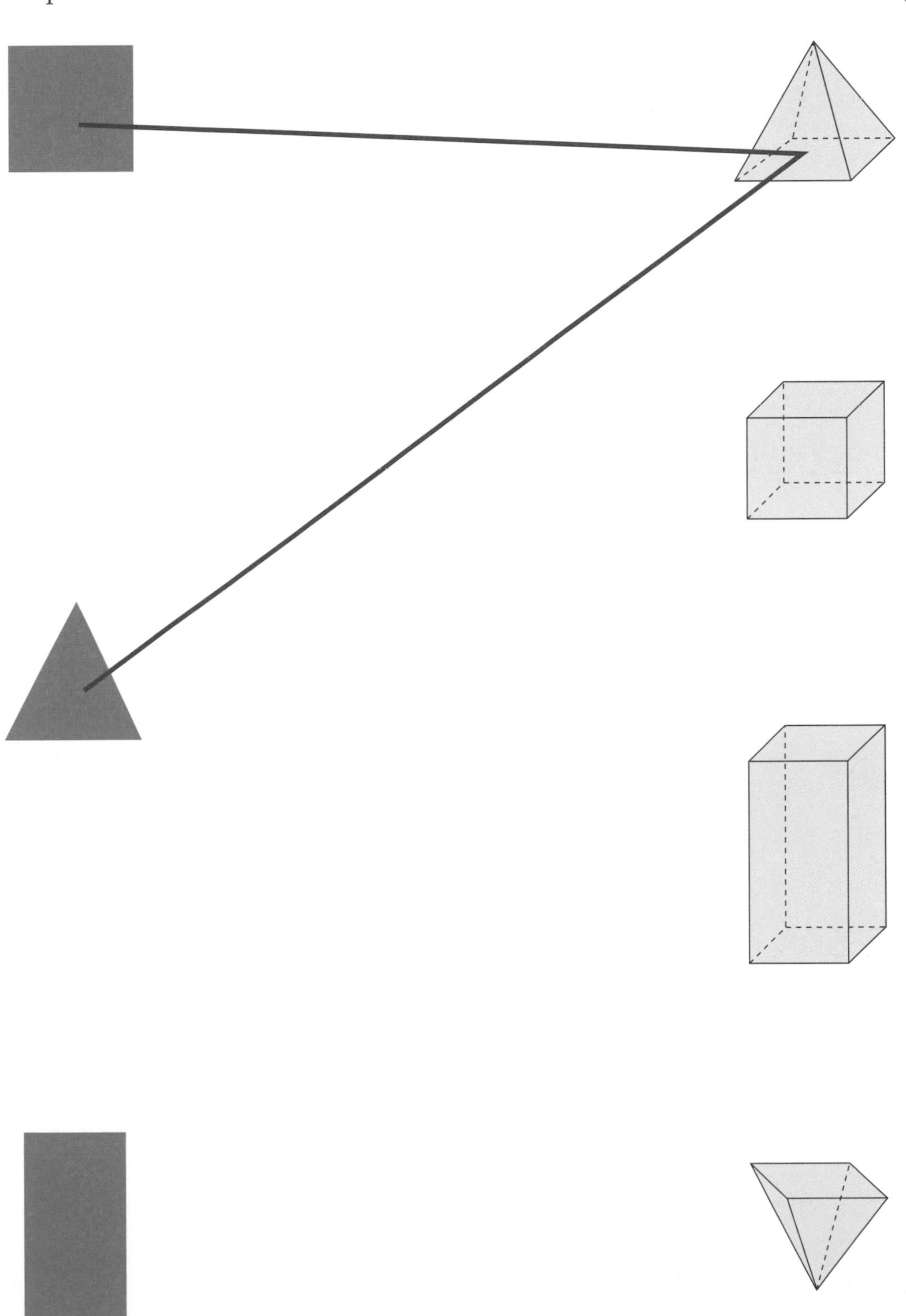

What's My Name?

WRITE the name of a solid shape that each person is describing.

I'm thinking of a solid shape that has five vertices.

1. ______________________

I'm thinking of a solid shape that has six faces.

2. ______________________

I'm thinking of a solid shape that has different shapes for faces.

3. ______________________

I'm thinking of a solid shape that has fewer faces than it does edges.

4. ______________________

What Comes Next?

DRAW the shape or picture that comes next in each pattern.

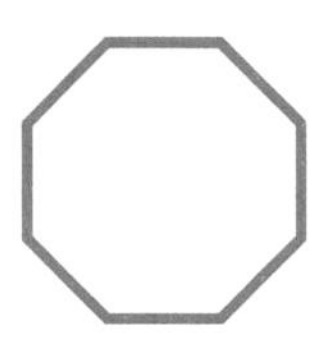

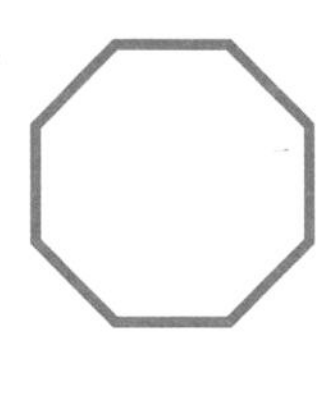

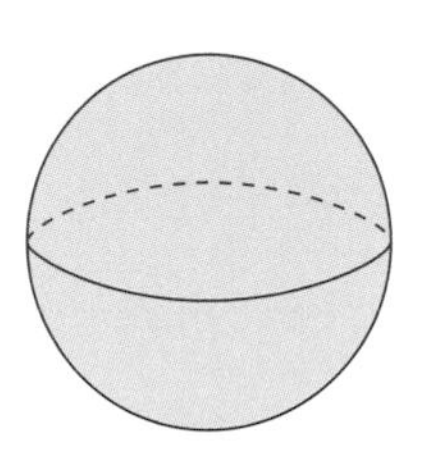
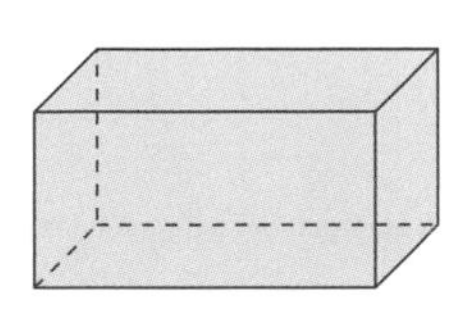
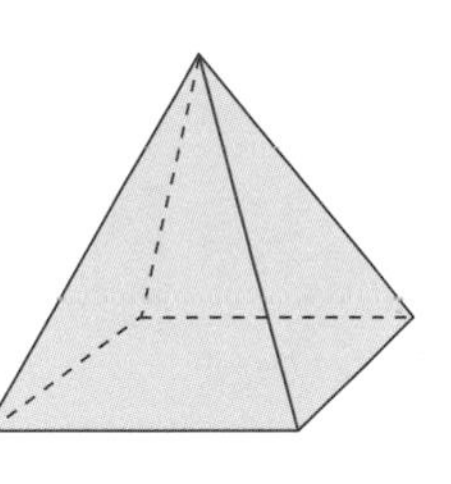
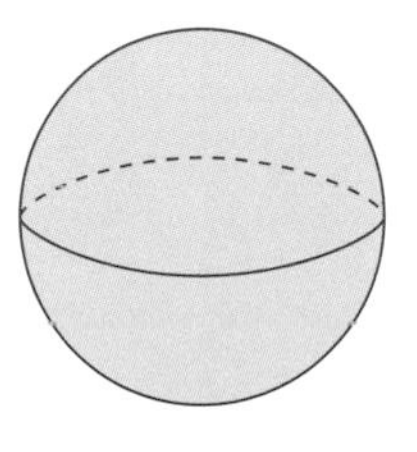
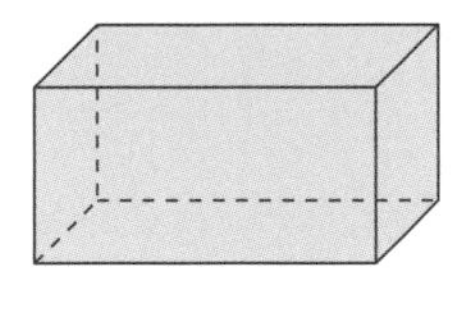
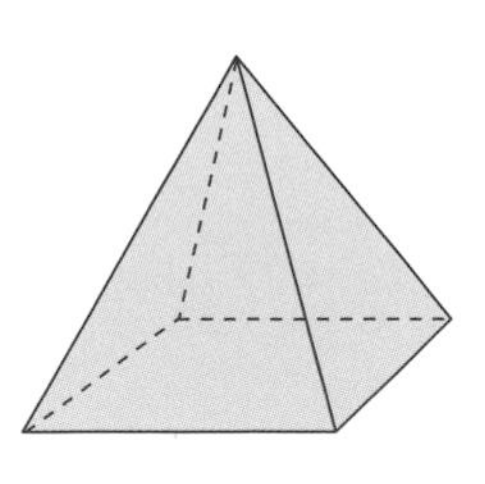

Spiraling Sequence

DRAW and COLOR the shapes to finish the spiral pattern.

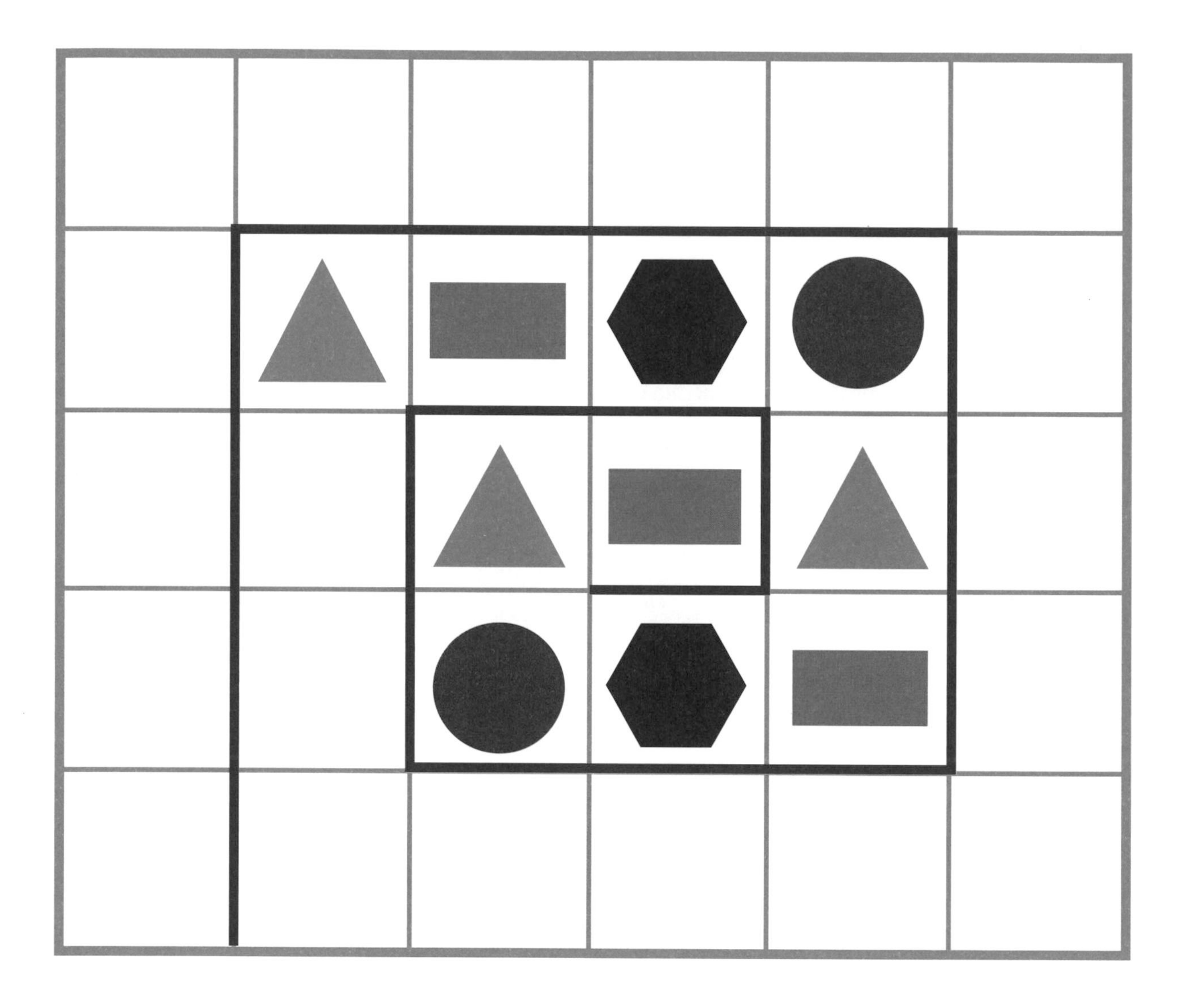

Puzzling Patterns

CIRCLE the set of shapes that completes the pattern.

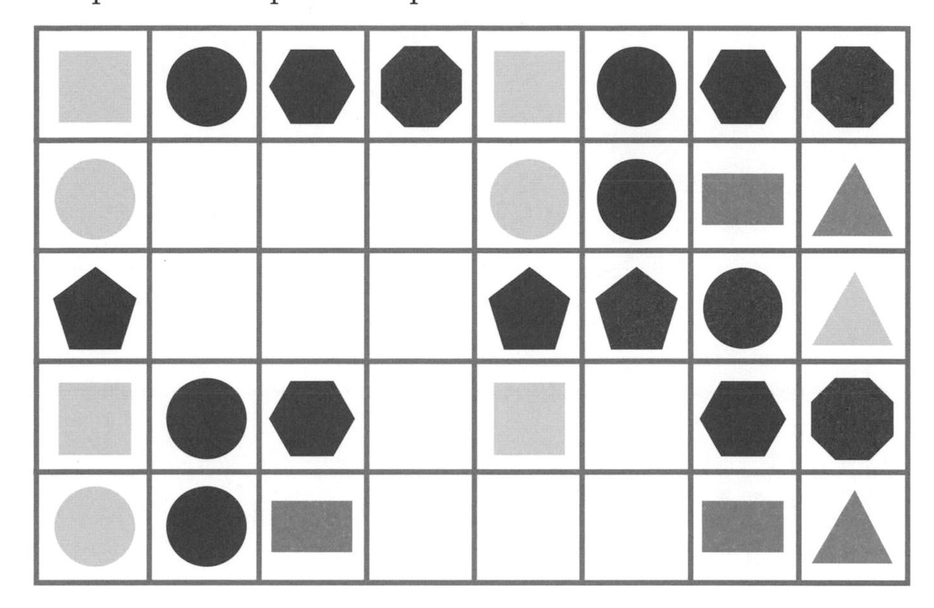

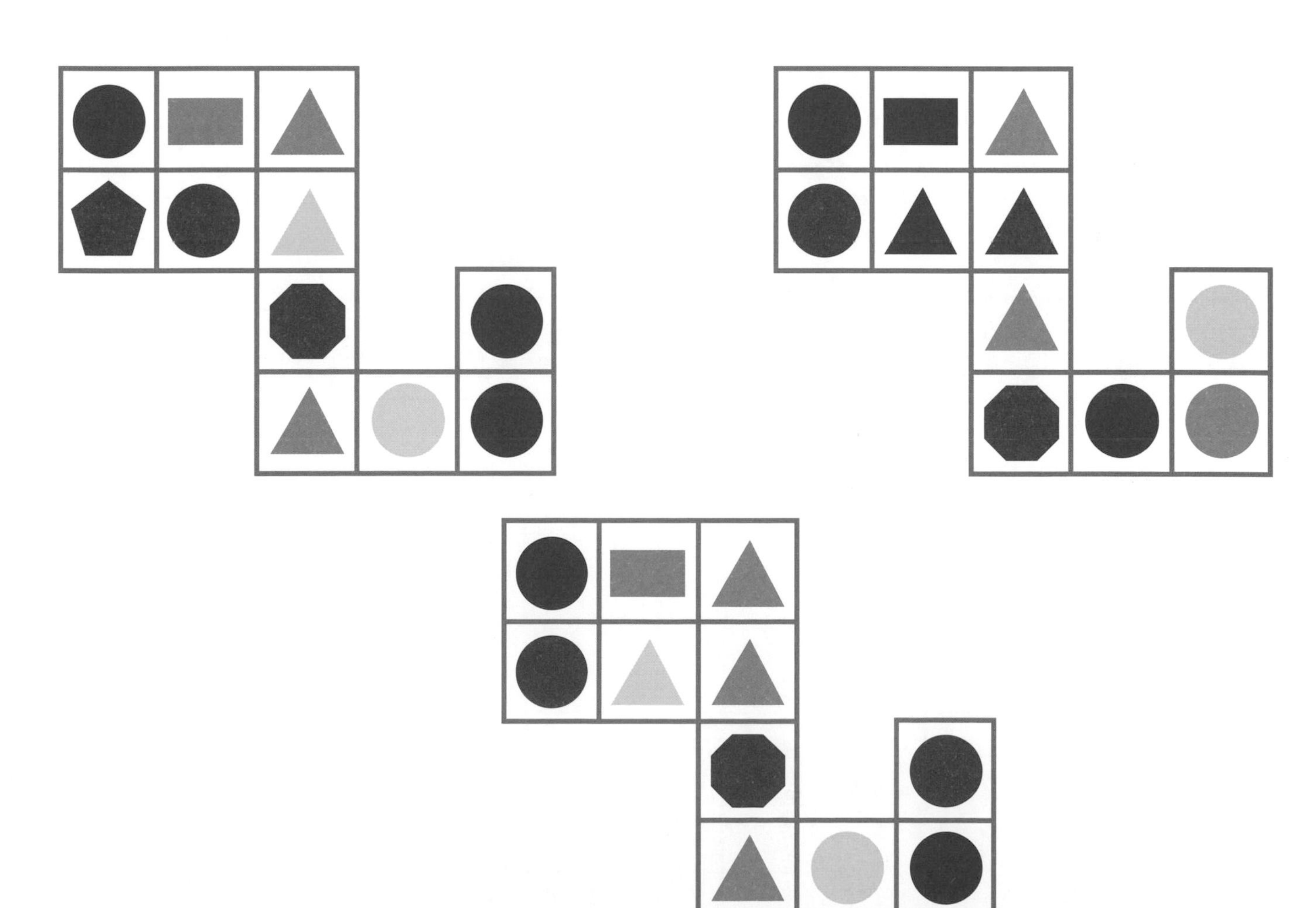

Puzzling Patterns

CIRCLE the set of shapes that completes the pattern.

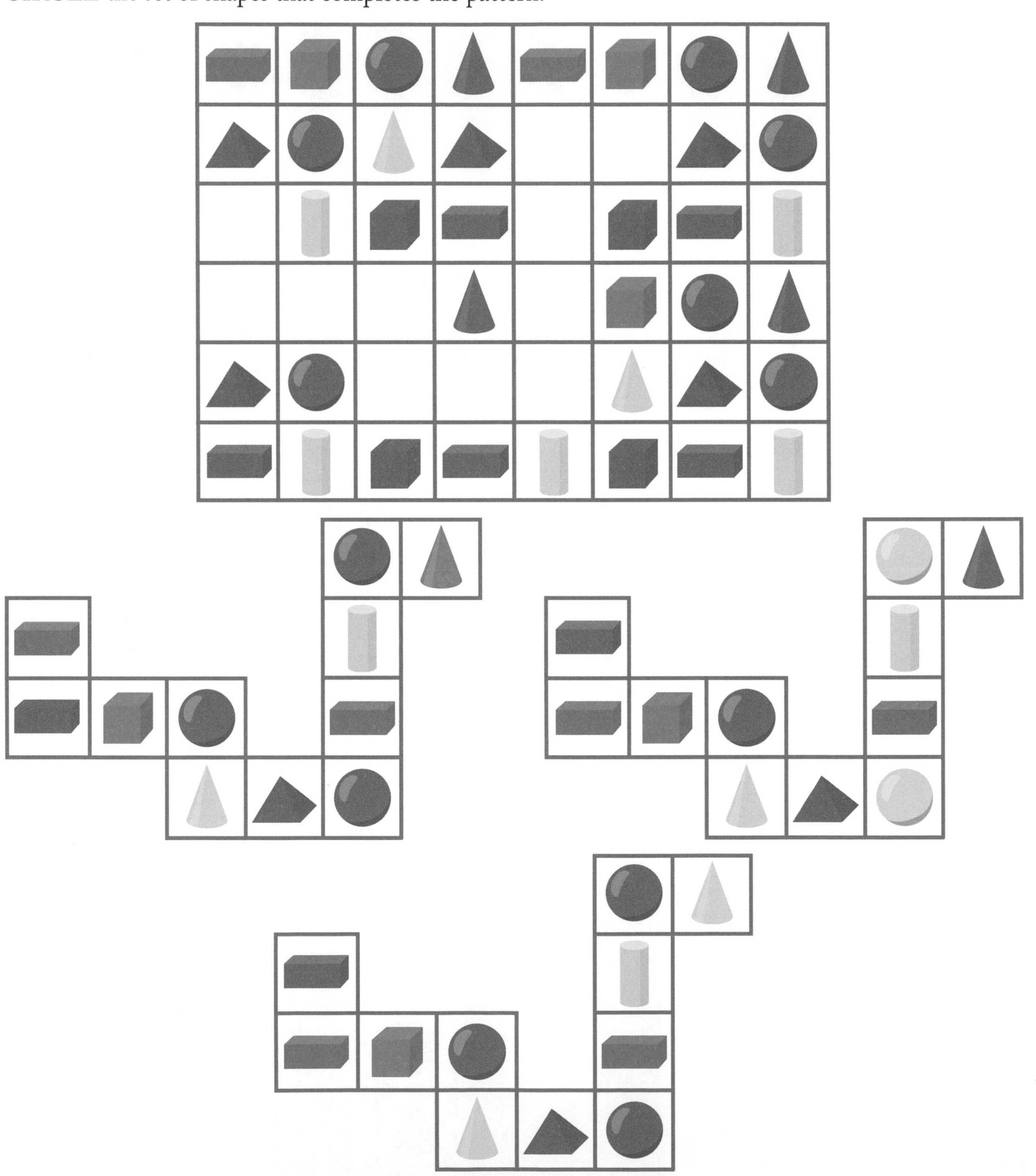

Card Tricks

CUT OUT these cards. Use these cards for page 49.

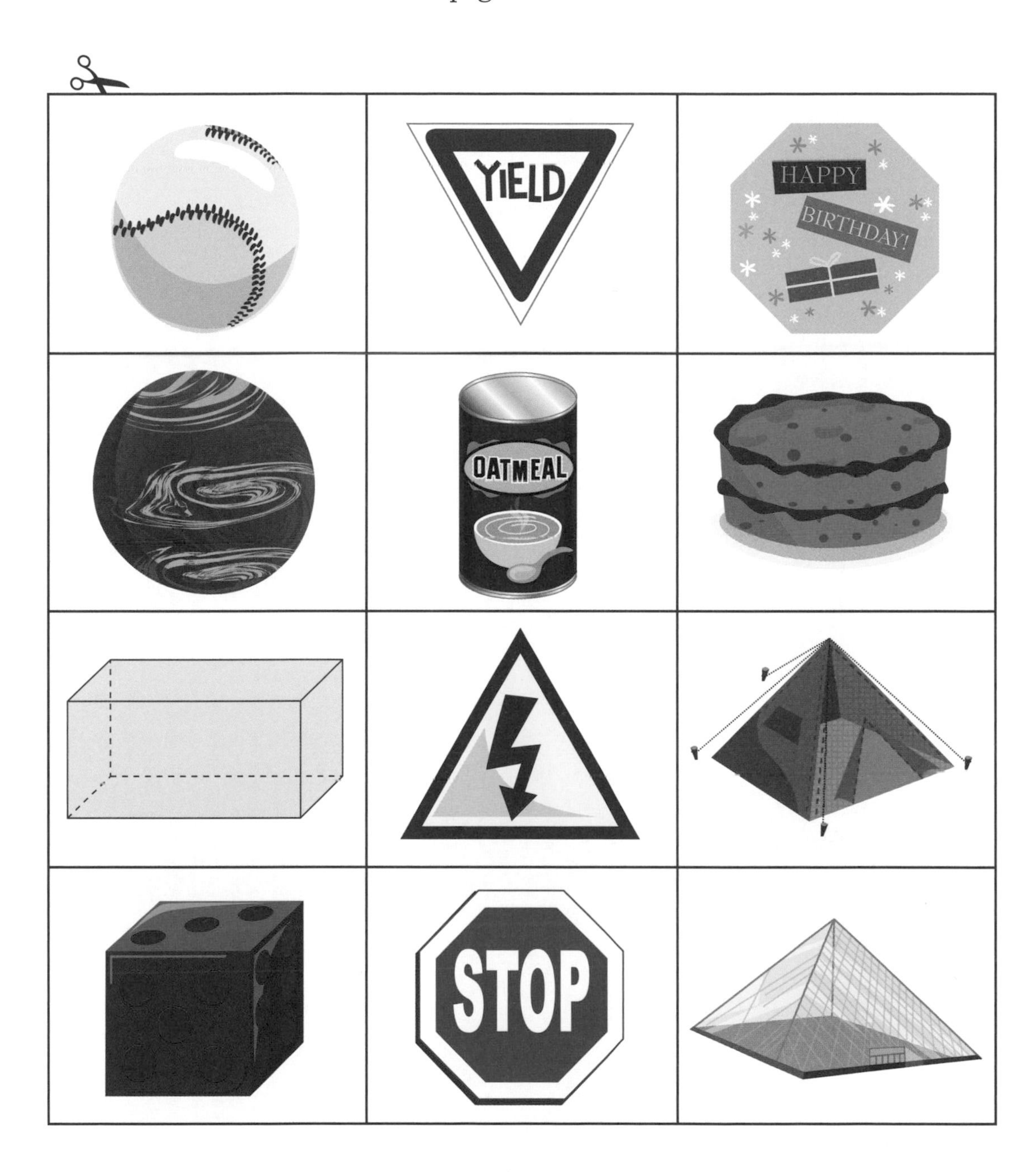

Card Tricks

Using the cards from page 47, PLACE two cards next to each picture to make a set. (Save the cards to use again).

HINT: Look for things that the shapes of each picture have in common.

Finish the Set

DRAW a shape that belongs in each set.

What Comes Next?

DRAW and COLOR the shape or picture that comes next in each pattern.

STOP

STOP

YIELD

ONE WAY

STOP

STOP

YIELD

ONE WAY

Spiraling Sequence

DRAW and COLOR the shapes to finish the spiral pattern.

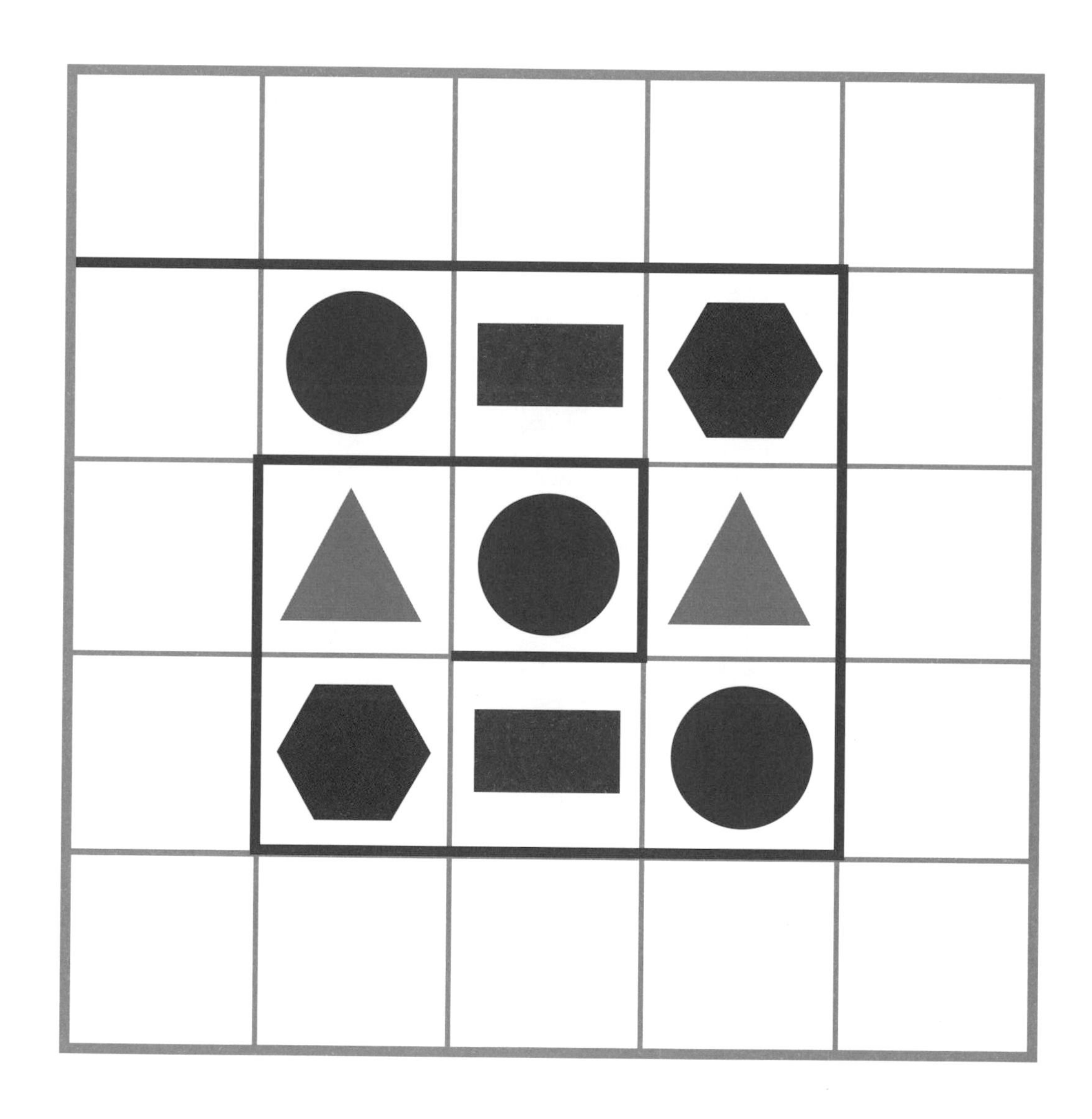

Card Tricks

Using the cards from page 47, PLACE all cards face down. CHOOSE two cards. If they're matching shapes, WRITE down the names of the shapes. If they don't match, TURN OVER both cards and try again. Continue until five matches have been found.

Match 1: ____________________ ____________________

Match 2: ____________________ ____________________

Match 3: ____________________ ____________________

Match 4: ____________________ ____________________

Match 5: ____________________ ____________________

Finish the Set

DRAW and COLOR two missing shapes to complete the pairs.

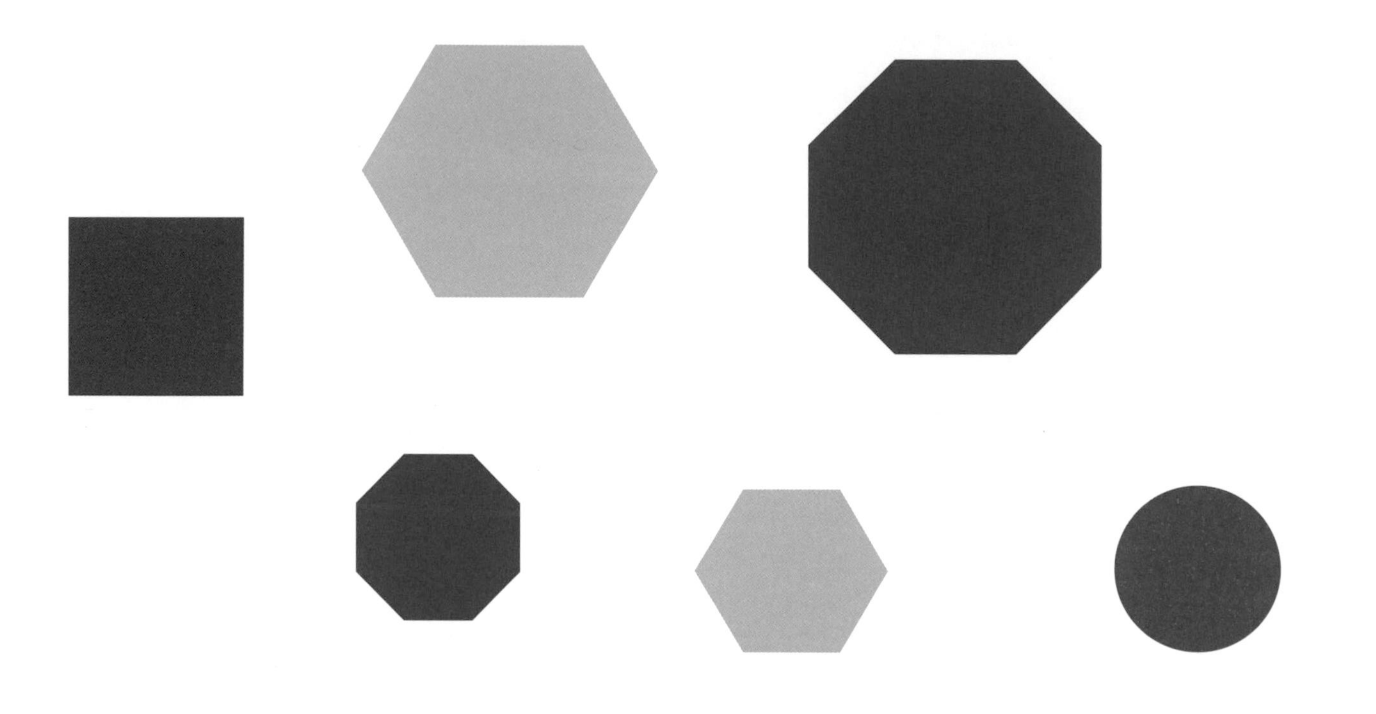

DRAW and COLOR two missing solid shapes to complete the pairs.

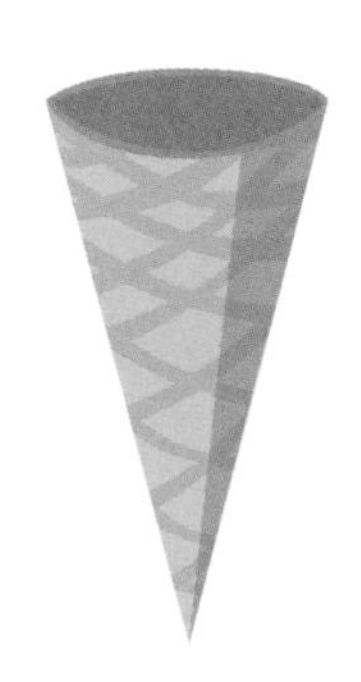

Mirror, Mirror

DRAW the line or lines of symmetry for each shape or picture.

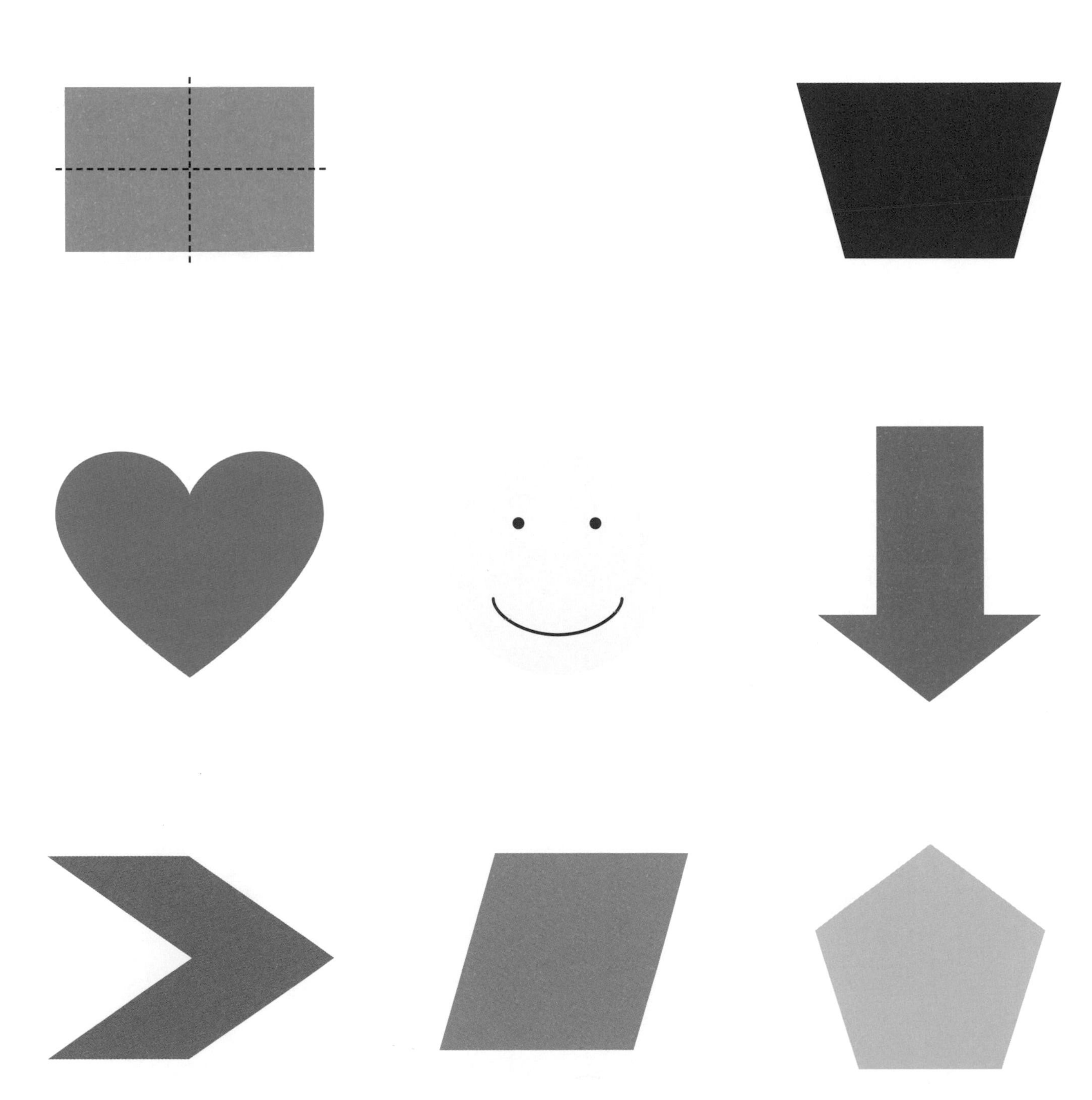

Mirror, Mirror

DRAW the line or lines of symmetry through each letter.

B

C

M

E

H

D

Picture Perfect

Which picture is symmetrical? CIRCLE the picture that is perfectly symmetrical.

Line Counter

DRAW all of the lines of symmetry in each shape or picture. Then WRITE the number of lines of symmetry for each.

1. ______

2. ______

3. ______

4. ______

5. ______

6. ______

All Lined Up

WRITE the number of lines of symmetry for each shape. Then DRAW an example of another shape with the same number of lines of symmetry.

	Number of Lines of Symmetry	Another Example

Half Again

Each object is symmetrical. DRAW the missing half of each object.

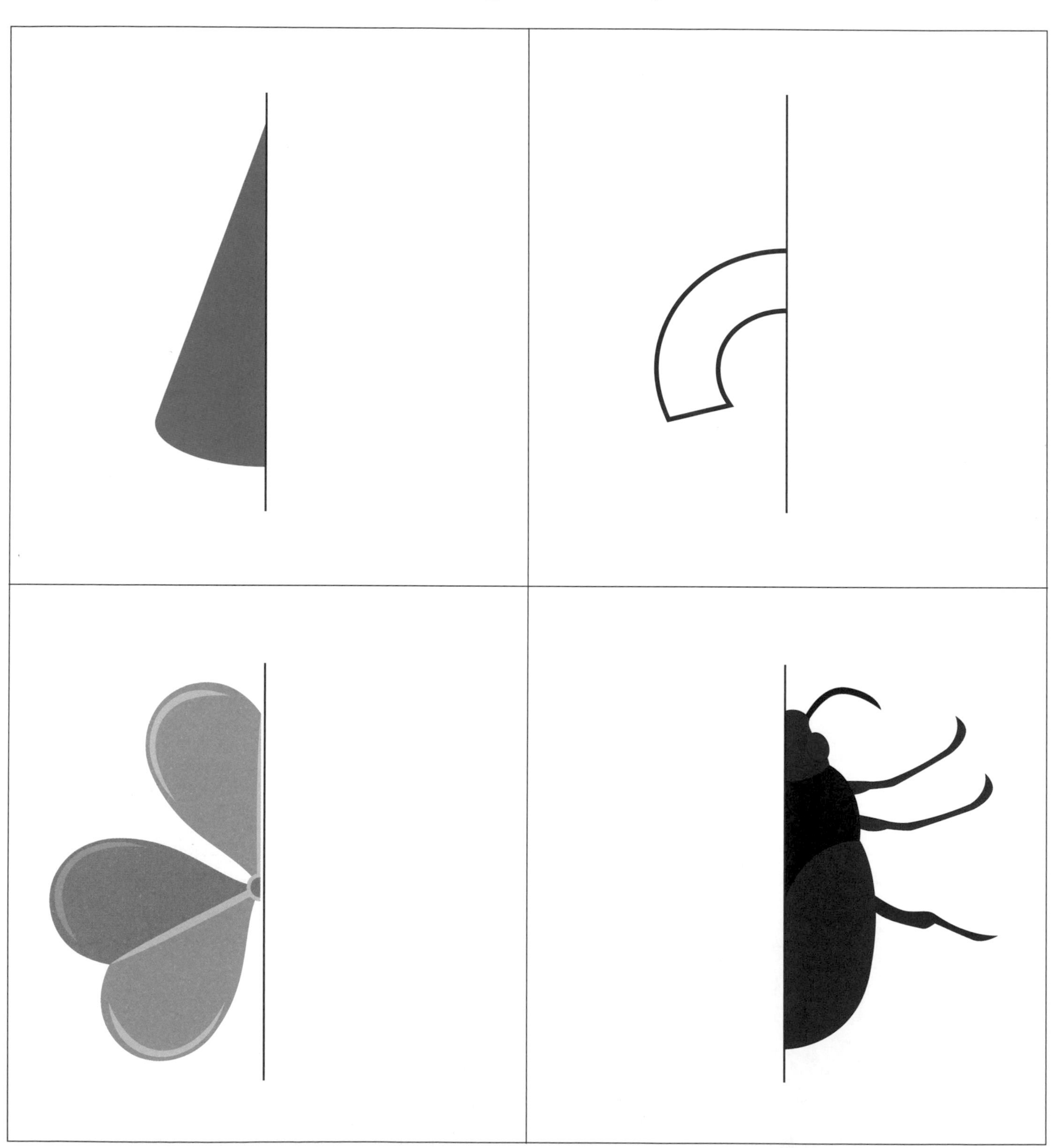

Color Flip

COLOR each design so it is symmetrical.

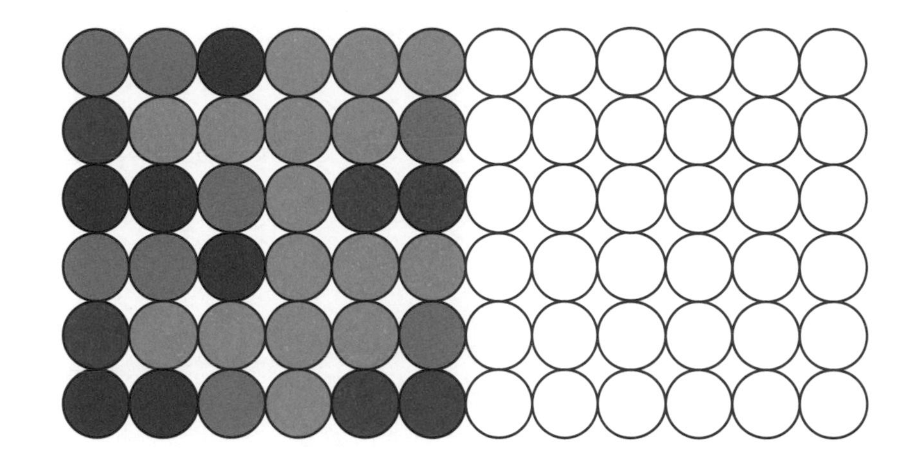

Color Flip

COLOR three different symmetrical designs.

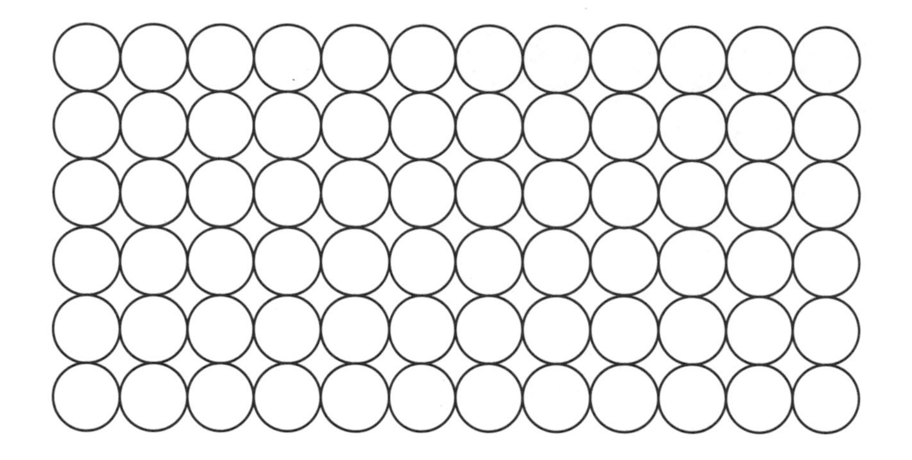

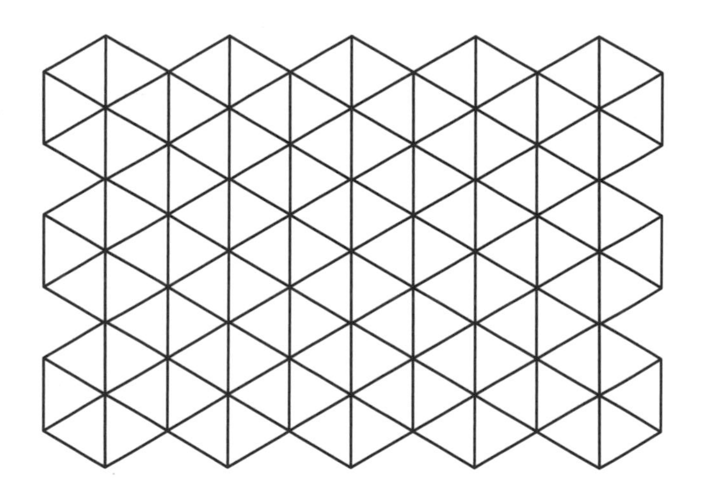

Mirror, Mirror

DRAW a line of symmetry through each picture. WRITE the number of lines of symmetry.

HINT: Each image may have no line, one line, or more than one line of symmetry.

1. ______

2. ______

3. ______

4. ______

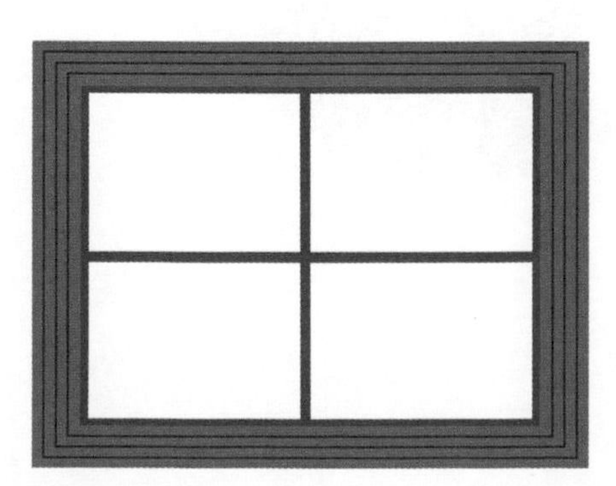

5. ______

6. ______

Picture Perfect

Which butterfly is symmetrical? CIRCLE the butterfly that is perfectly symmetrical.

Half Again

Each object is symmetrical. DRAW the missing half.

Cool Kaleidoscope

DRAW a line of symmetry. COLOR the design so it is symmetrical.

Any Which Way

A **flip**, **slide**, or **turn** has been applied to each shape. WRITE *flip*, *slide*, or *turn*.
HINT: Some may have more than one possible answer.

Example:

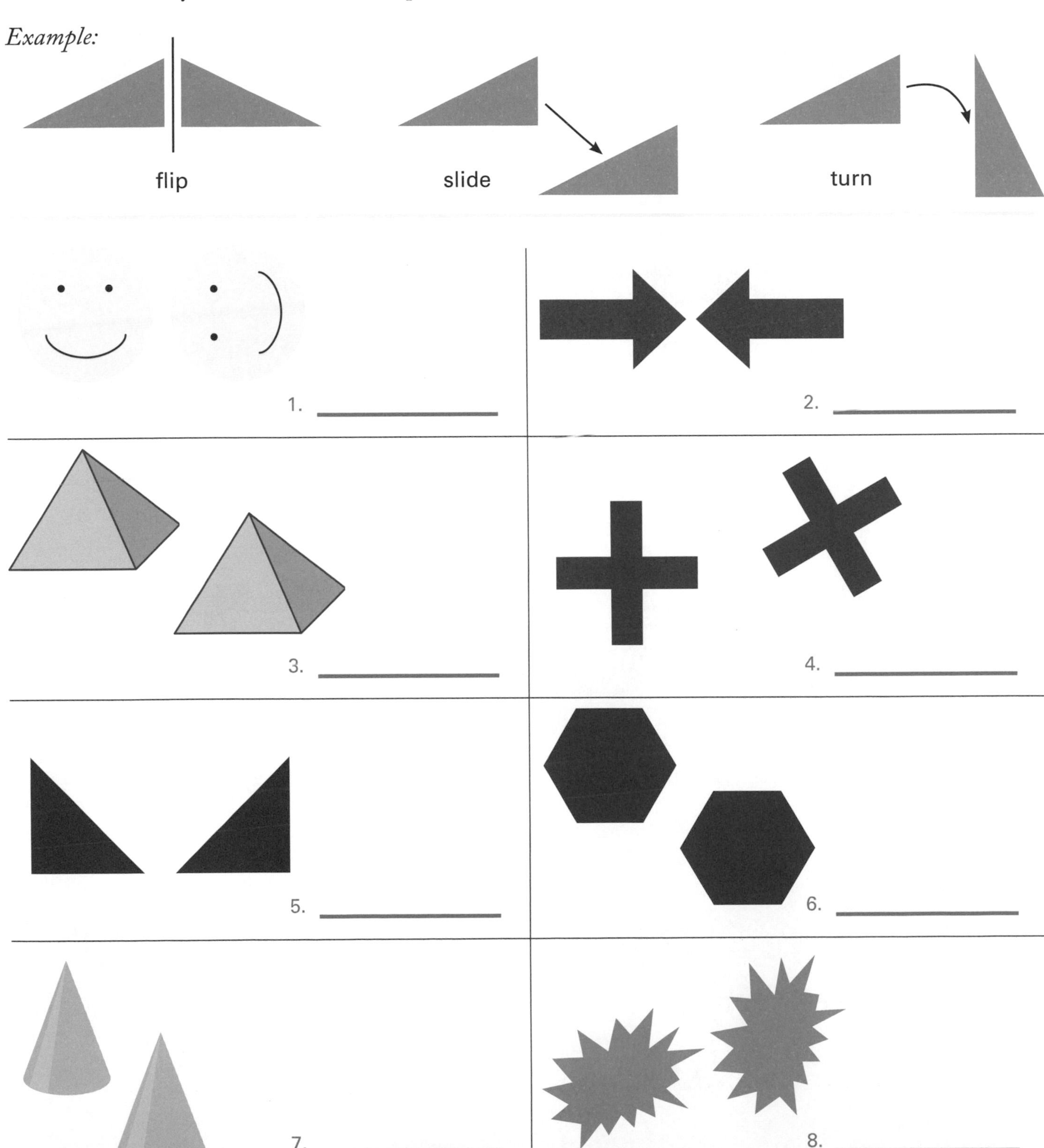

Shifting Shapes

DETERMINE if a flip has been applied to each shape. CIRCLE the pairs of shapes that demonstrate a flip.

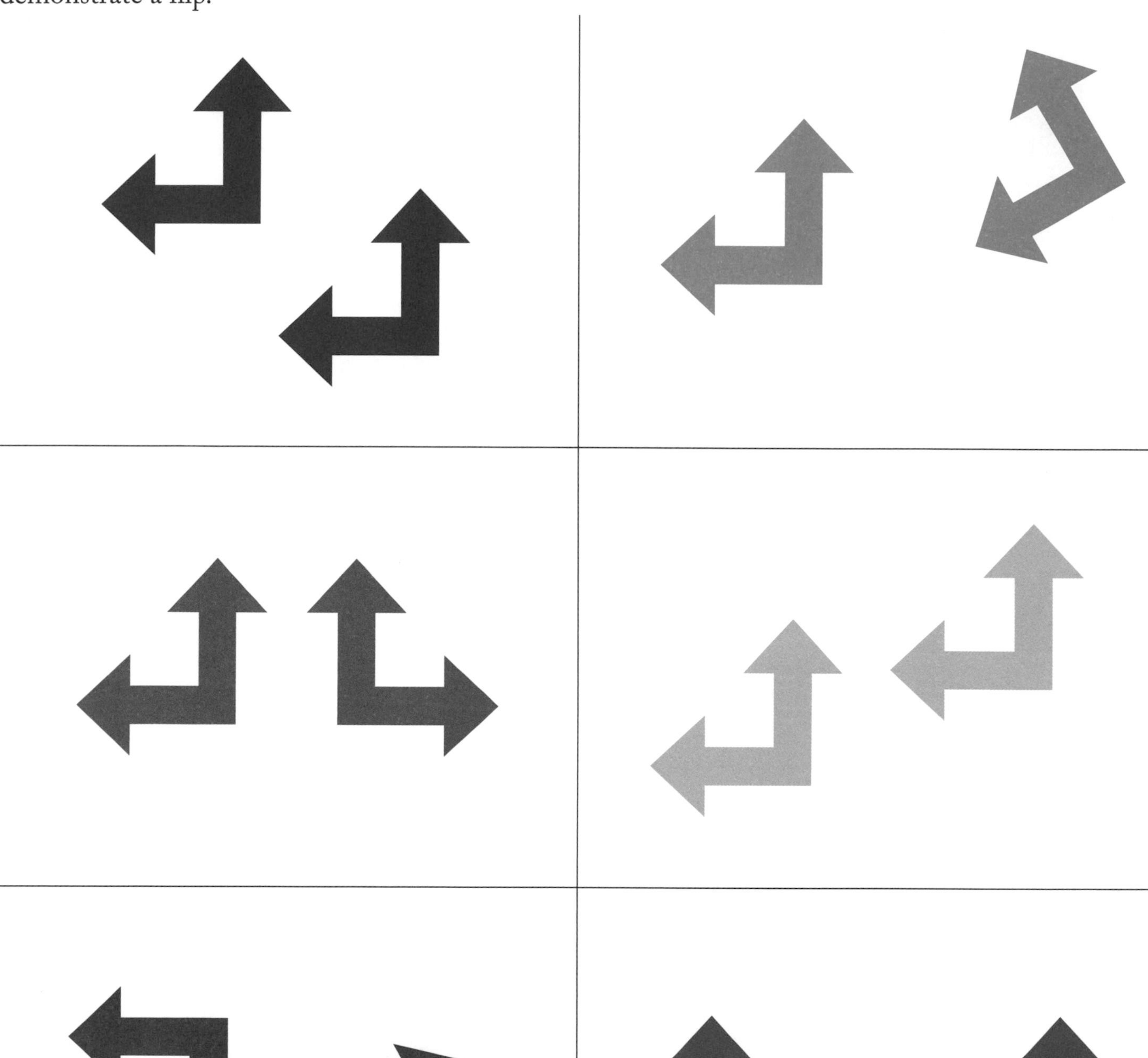

Shifting Shapes

DETERMINE if a slide has been applied to each shape. CIRCLE the pairs of shapes that demonstrate a slide.

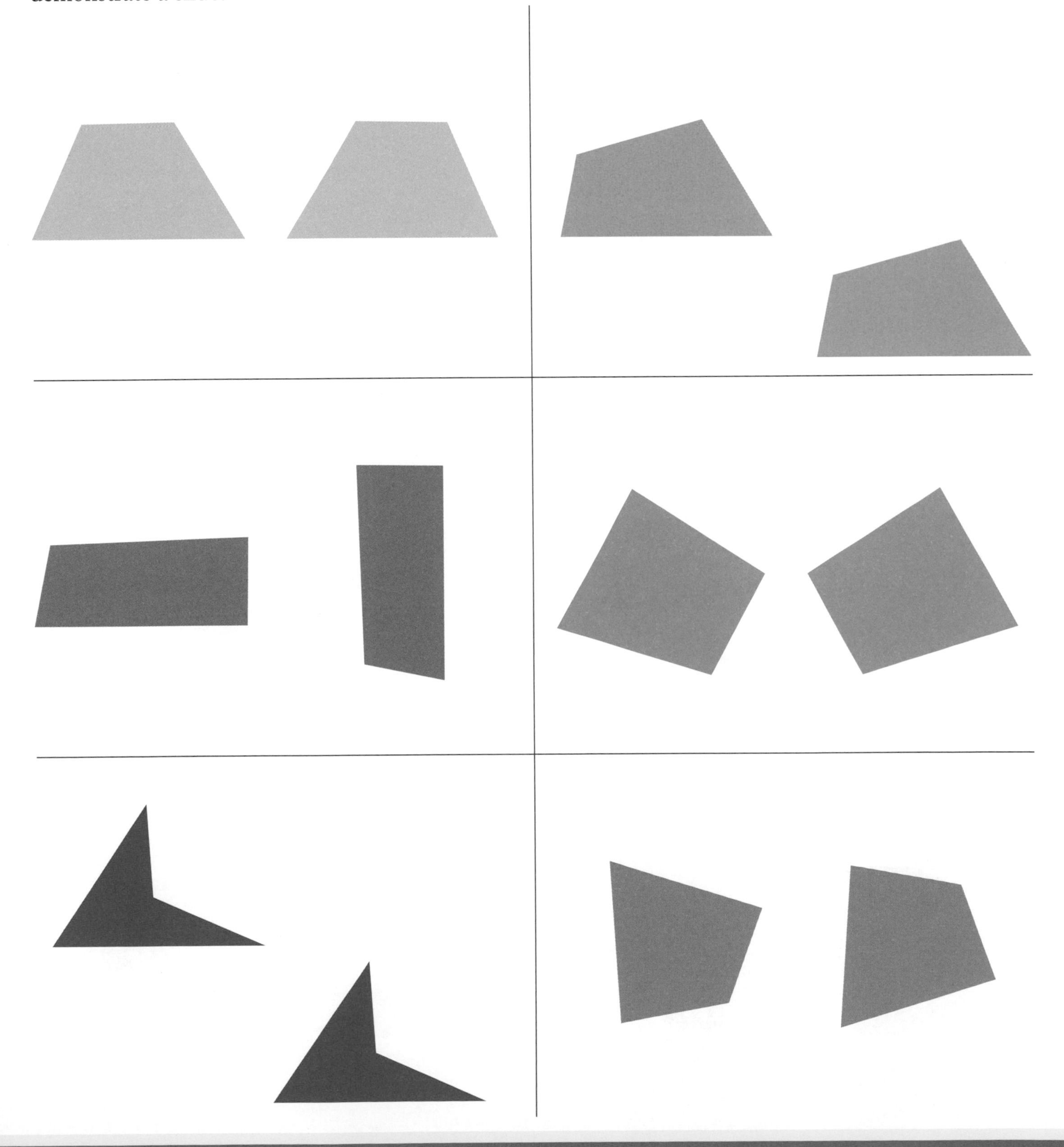

Shifting Shapes

DETERMINE if a turn has been applied to each shape. CIRCLE the pairs of shapes that demonstrate a turn.

Match Up

DRAW a line to match each pair of shapes to the correct description.

HINT: Each description can be used more than once.

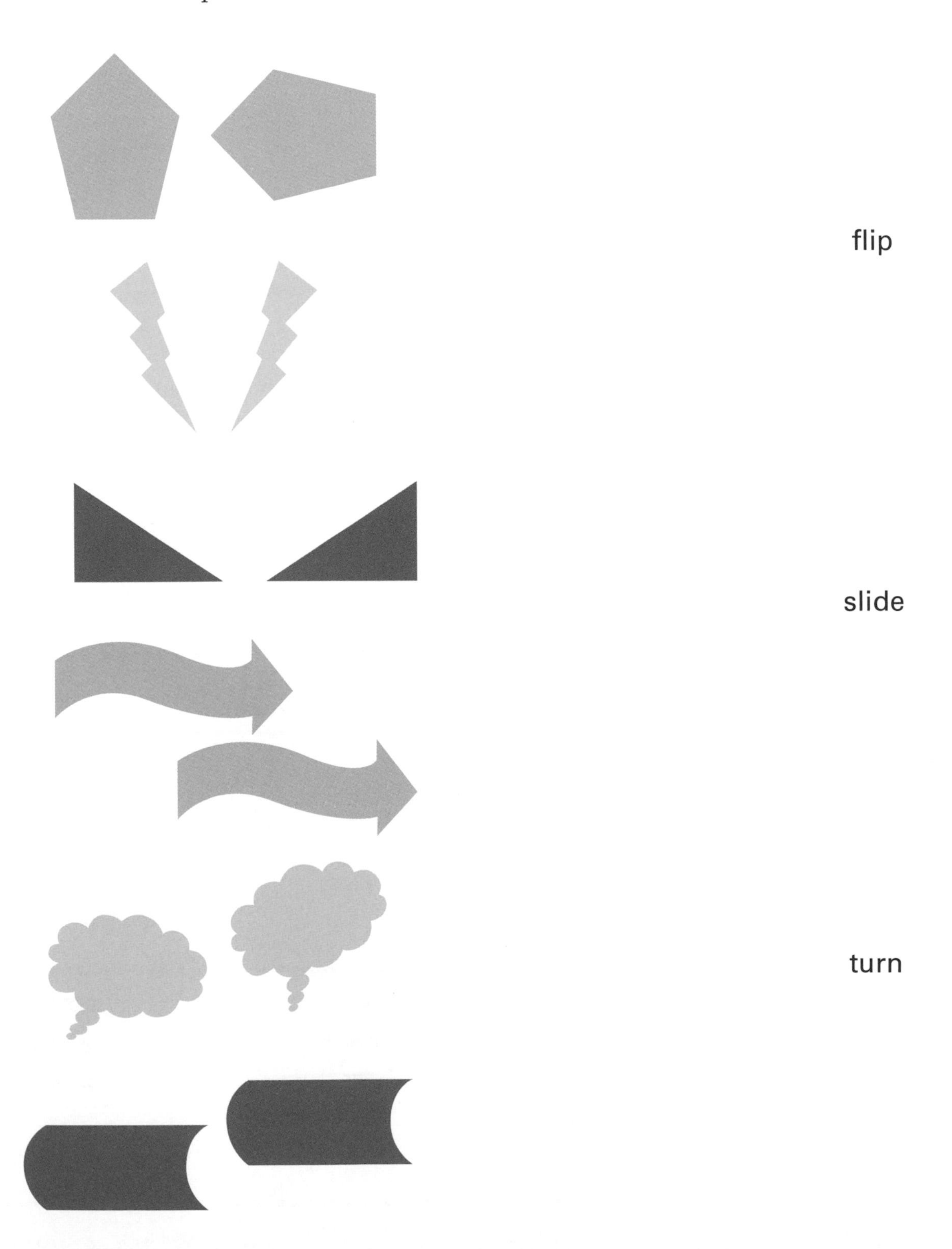

Which One?

PLACE an X on the object in each row that is **not** a flip, slide, or turn of the original object.

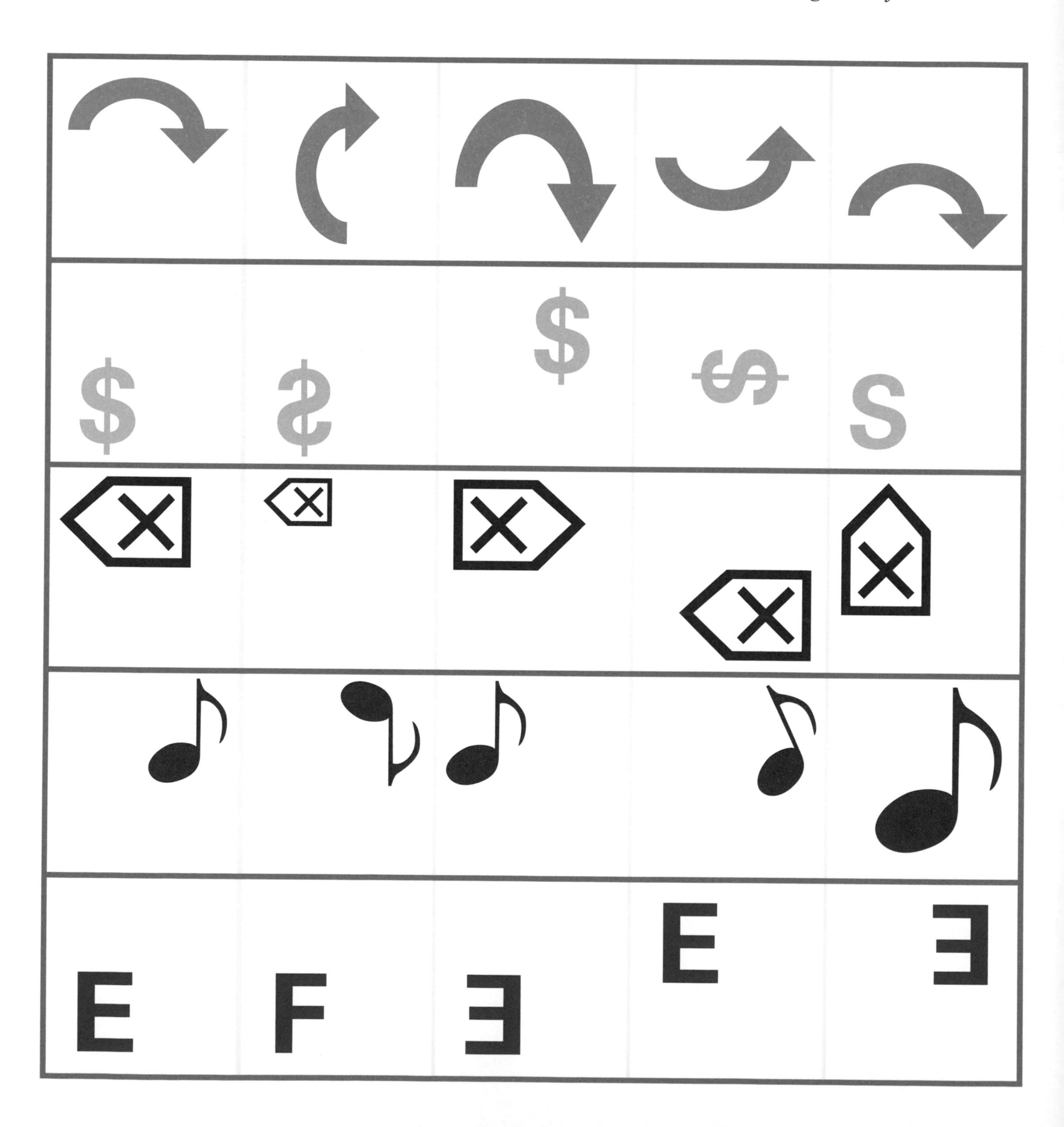

Shape Creator

DRAW a shape on the first grid. Then APPLY the stated transformation on the other grids.

shape

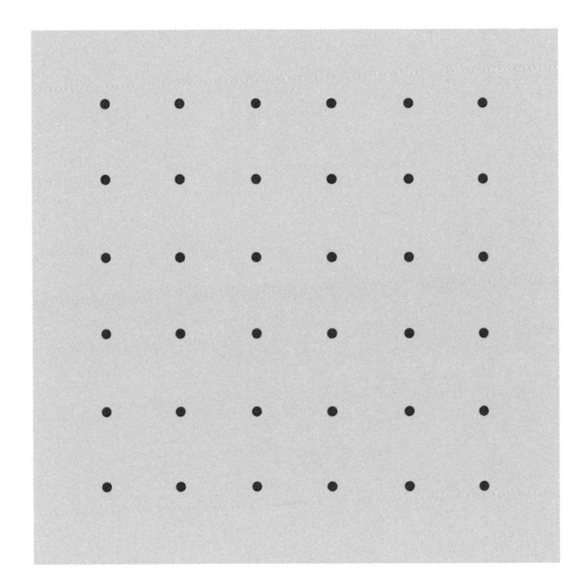

slide

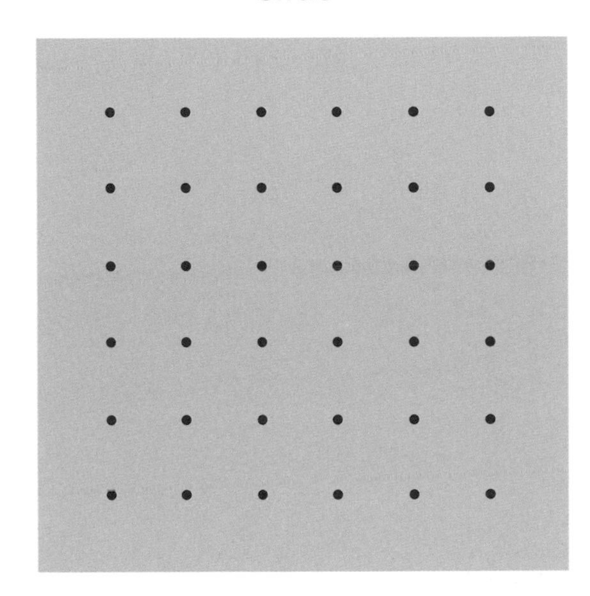

turn

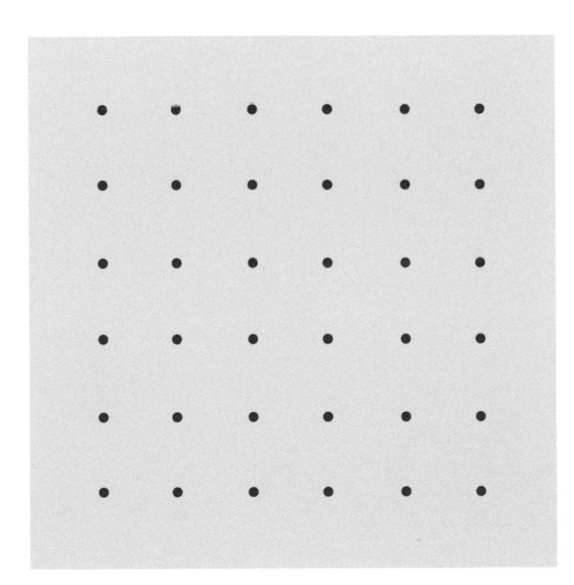

flip

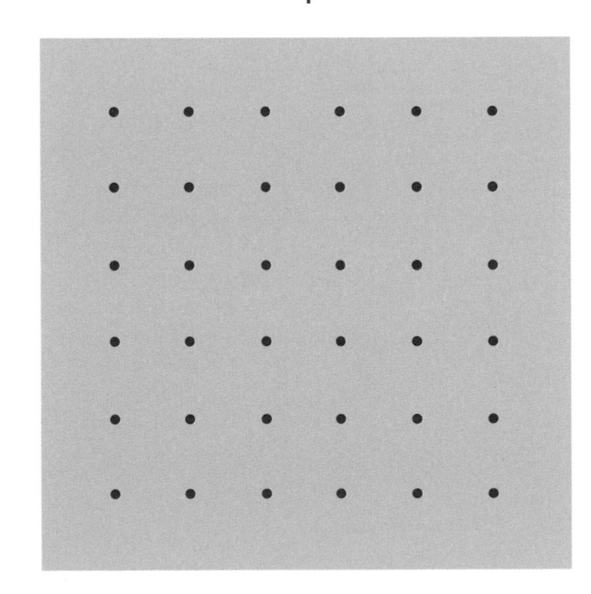

Puzzling Pentominoes

Using the pentomino pieces from page 115, TRACE and COLOR one piece on this page. Then APPLY a flip, slide, and turn to the shape. TRACE and COLOR the result. (Save the pieces to use again.)

Puzzling Pentominoes

Using the pentomino pieces from page 115, TRACE and COLOR one piece on this page. Then APPLY a flip, slide, and turn to the shape. TRACE and COLOR the result. (Save the pieces to use again.)

Flip and Turn

COLOR each shape to show the first shape flipped and turned.

	Flip	Turn

Flip and Turn

DRAW and COLOR each flag flipped and turned.

	Flip	Turn

Shape Creator

DRAW a shape on the first grid. Then APPLY the stated transformation on the other grids.

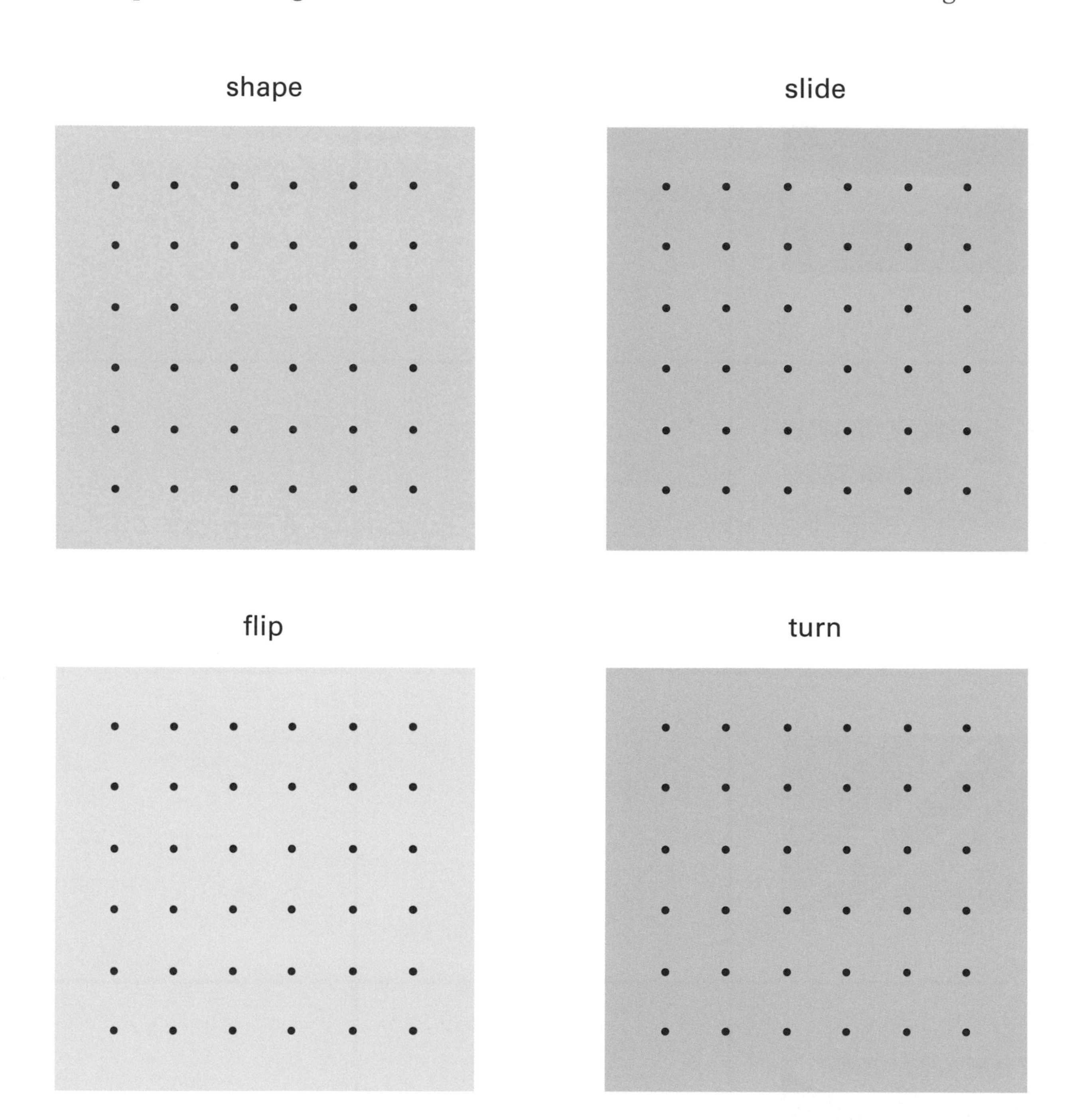

Any Which Way

WRITE *flip*, *slide*, or *turn*.

1. ____________

2. ____________

3. ____________

4. ____________

5. ____________

6. ____________

Which One?

PLACE an X on the object in each row that is **not** a flip, slide, or turn of the original object.

Z	Z	Z	N	Z
MOM	MOM	MOM	MOM	MOM

Shape Creator

DRAW and COLOR a shape in each grid on the left. Then DRAW and COLOR a flip, slide, or turn of the shape.

flip

slide

turn

Flip and Turn

DRAW the image flipped and turned.

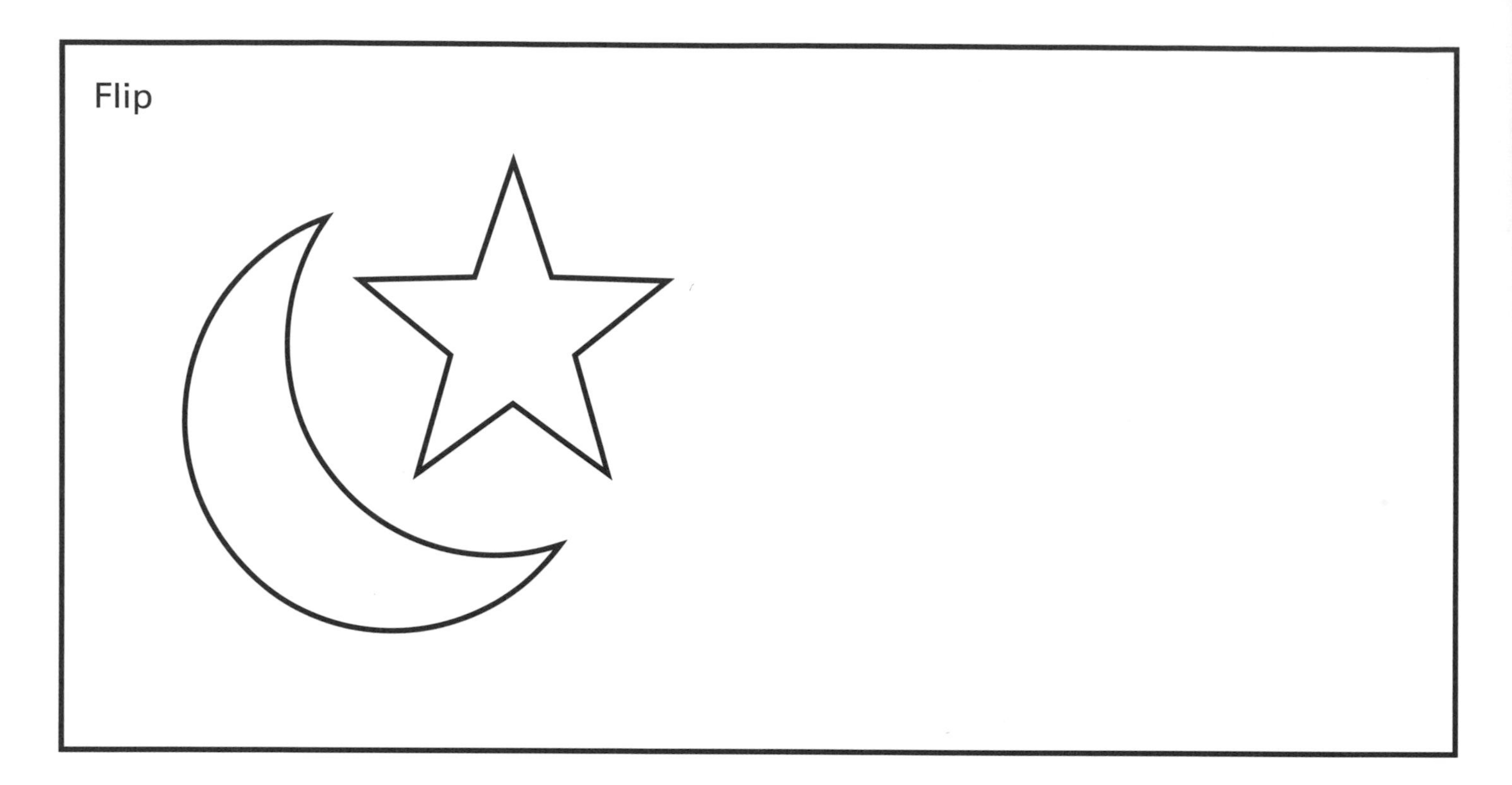

Around We Go

The distance around a two-dimensional shape is the **perimeter**. Add the lengths of each side to find the perimeter. WRITE the perimeter of each shape.

HINT: For shapes with sides that are all the same length, multiply the length of one side by the number of sides.

Example:

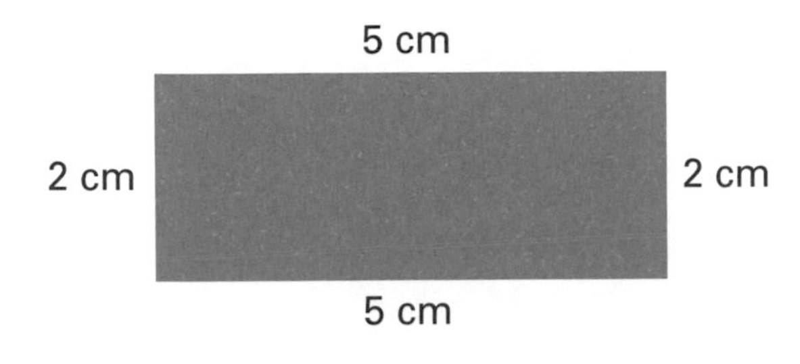

$2 + 5 + 2 + 5 = 14$
The perimeter of the rectangle is 14 centimeters (cm).

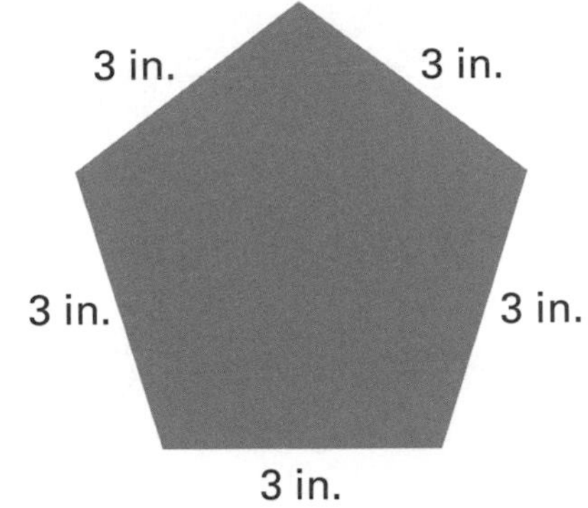

$3 \times 5 = 15$
The perimeter of the shape is 15 inches (in.).

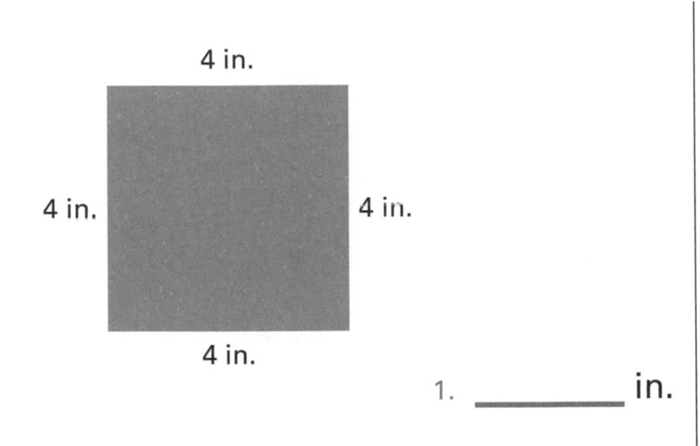

1. _______ in.

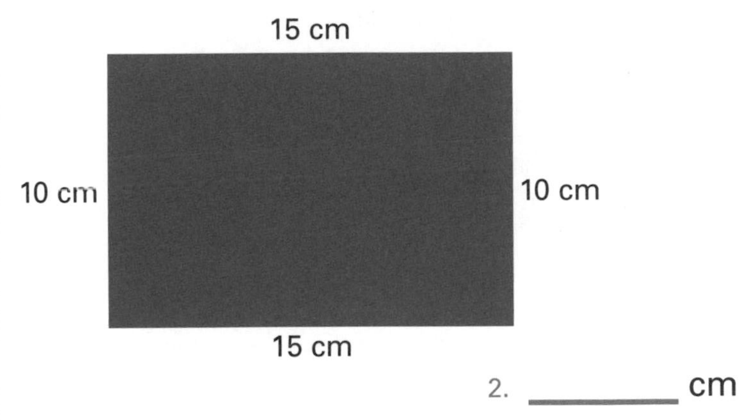

2. _______ cm

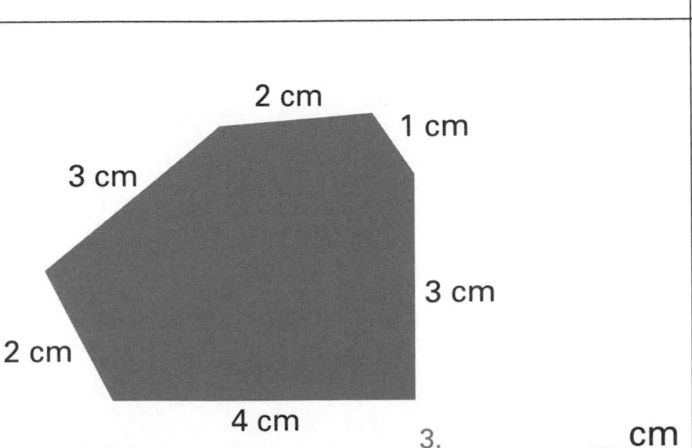

3. _______ cm

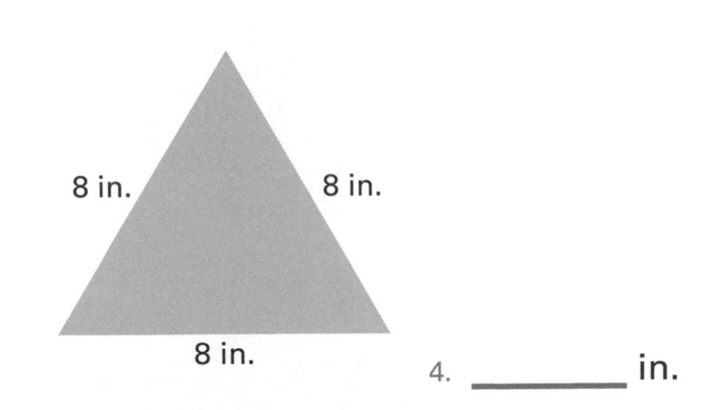

4. _______ in.

Around We Go

WRITE the perimeter of each shape.

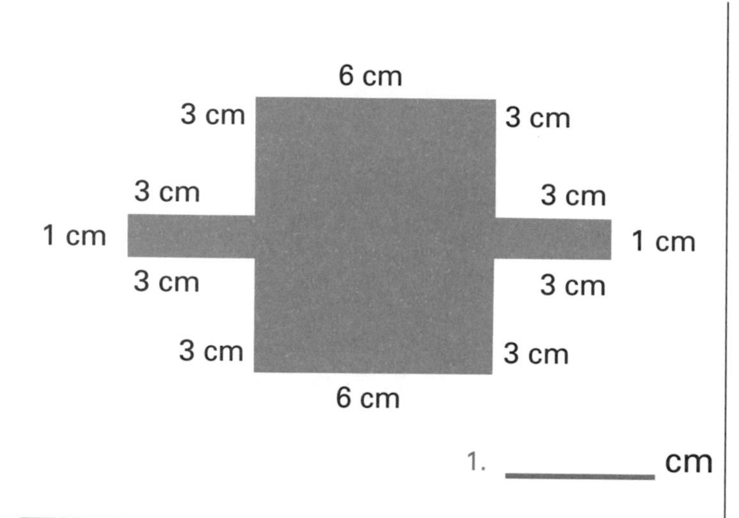

1. ________ cm

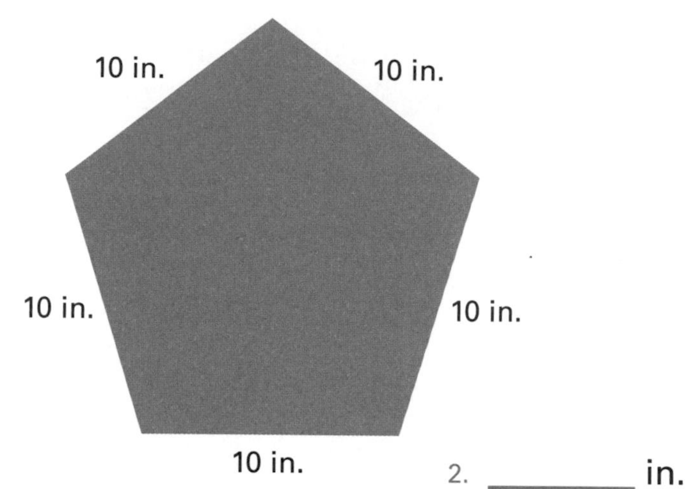

2. ________ in.

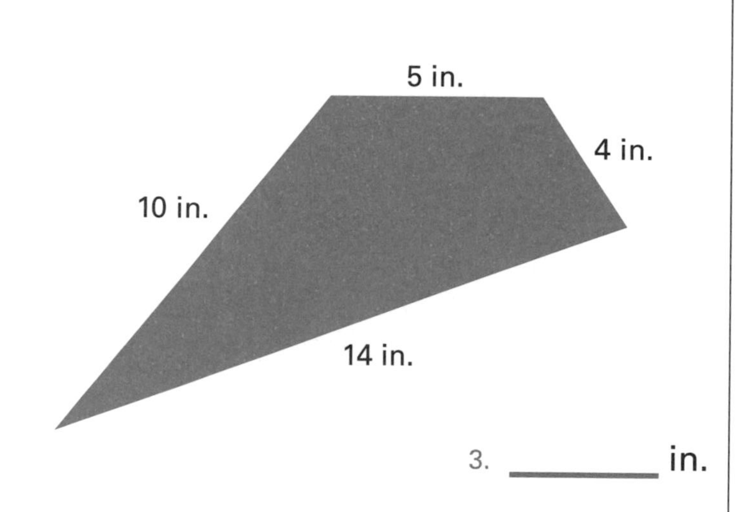

3. ________ in.

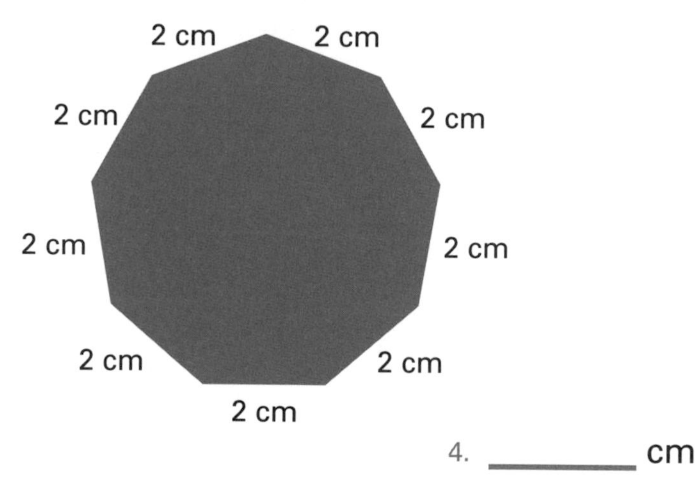

4. ________ cm

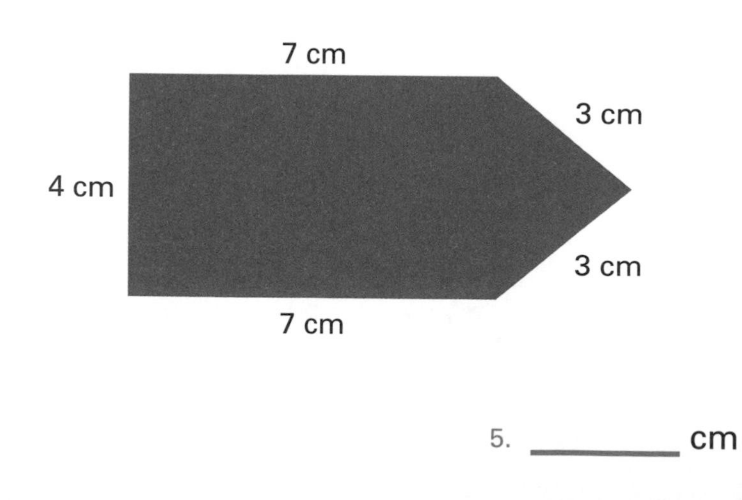

5. ________ cm

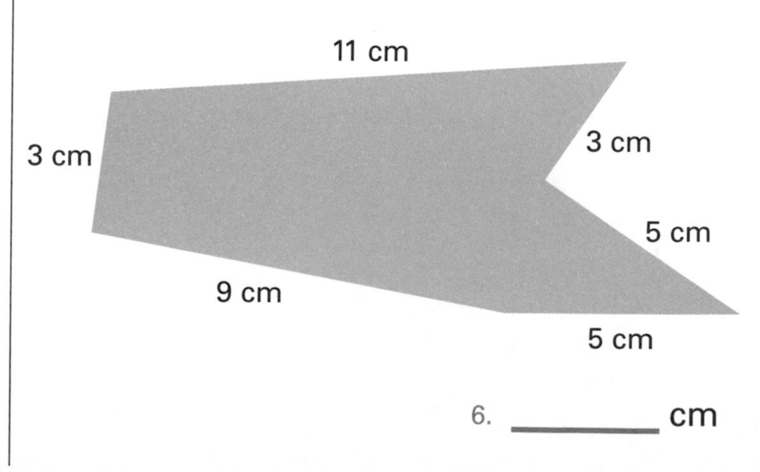

6. ________ cm

Shape Creator

Each dot is 1 centimeter away from the dot above, below, or beside it. DRAW four different shapes that each have a perimeter of 20 centimeters.

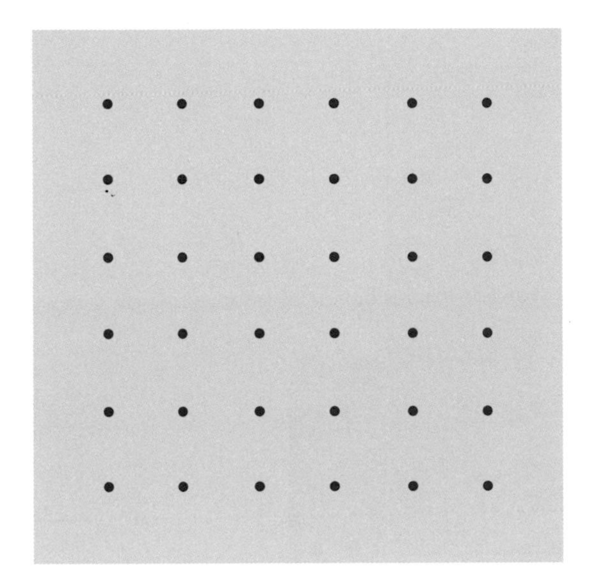

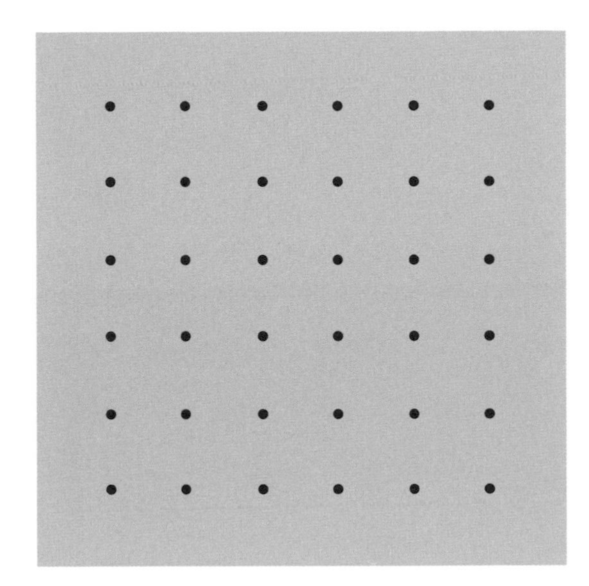

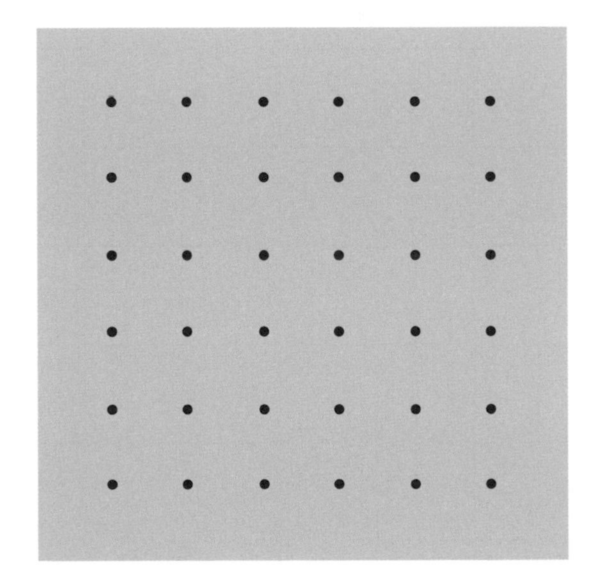

Puzzling Pentominoes

Using the pentomino pieces from page 115, PLACE the pieces to create a shape with the perimeter given without overlapping any pieces. TRACE and COLOR the shape created with the pentomino pieces. (Save the pentomino pieces to use again.)

Example:

Create a shape with a perimeter of 20 units.

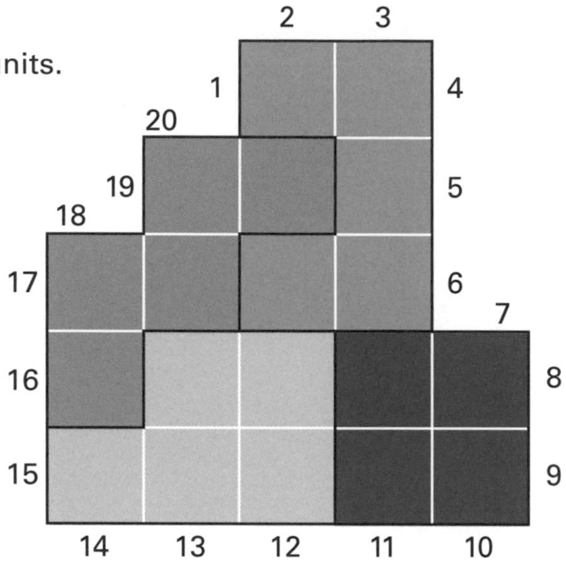

perimeter = 20 units

24 units

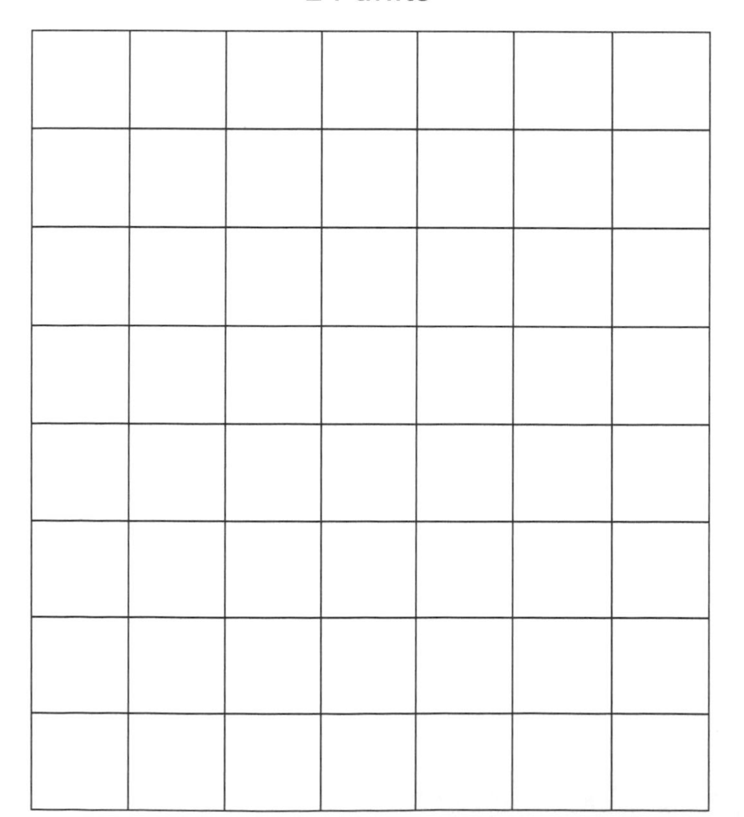

32 units

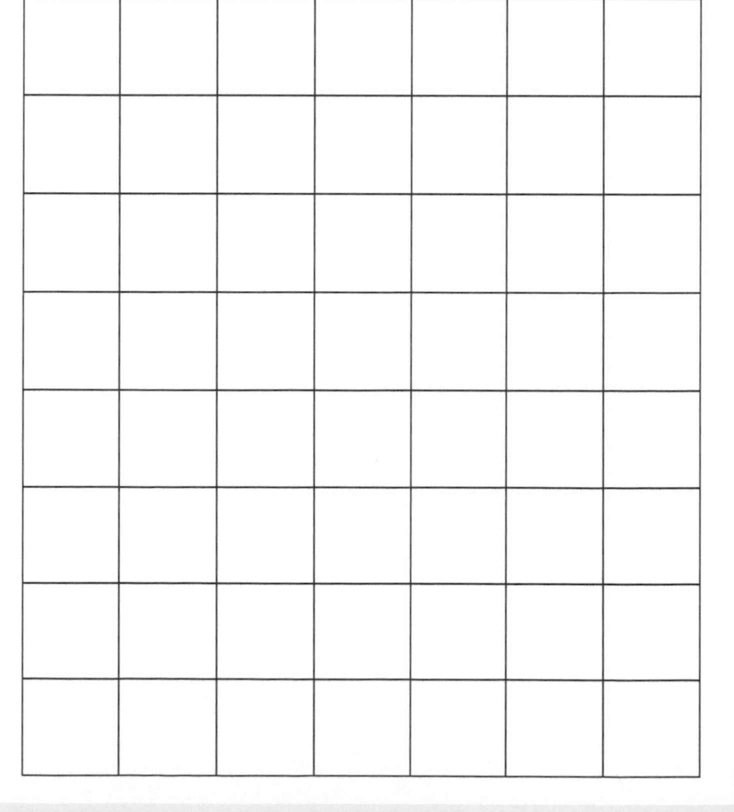

Puzzling Pentominoes

Using the pentomino pieces from page 115, PLACE the pieces to completely fill each shape without overlapping any pieces. Then WRITE the perimeter of each shape. (Save the pentomino pieces to use again.)

1. ________ units

2. ________ units

3. ________ units

Put It Together

Using the measurements, WRITE the perimeter of each colored section.

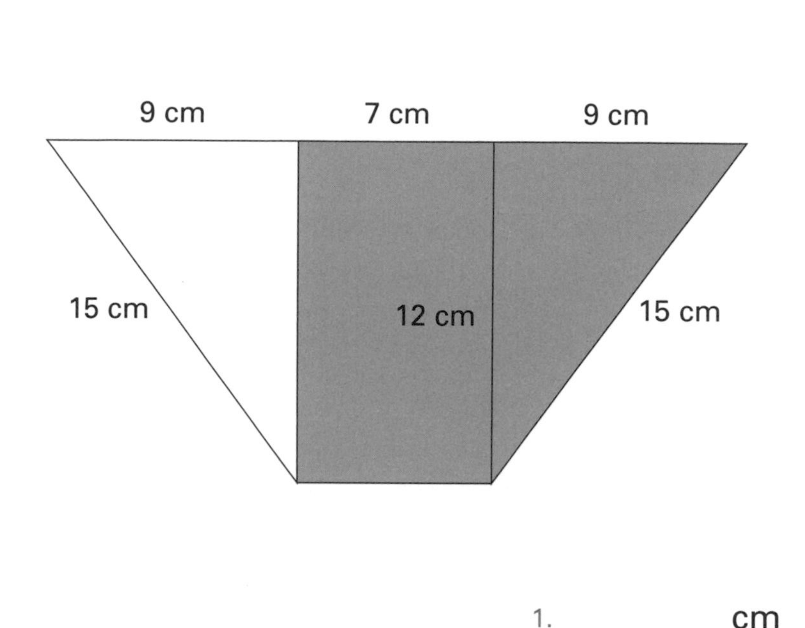

1. ________ cm

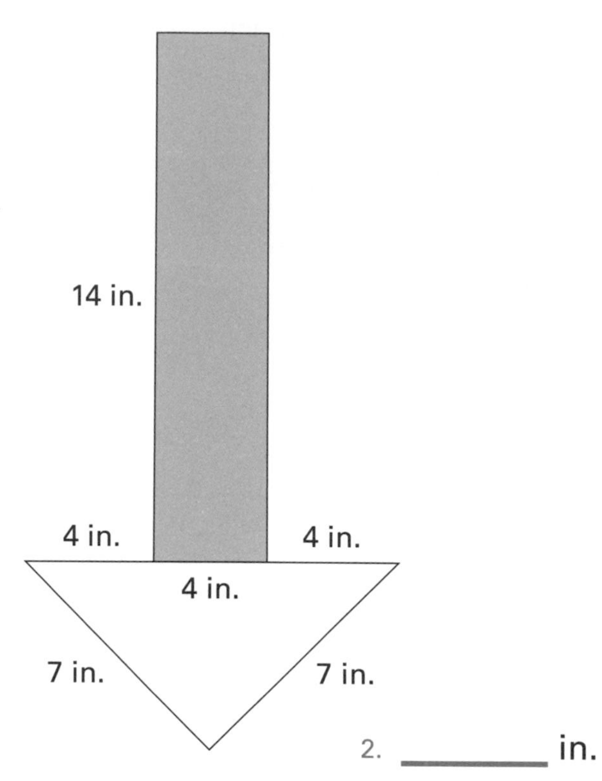

2. ________ in.

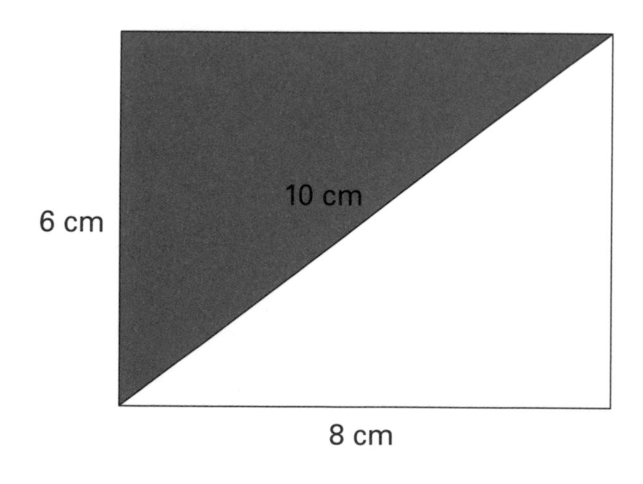

3. ________ cm

3 cm
3 cm
11 cm
3 cm
3 cm

4. ________ cm

Squared Away

Area is the size of the surface of a shape, and it is measured in square units. WRITE the area of each shape.

Example: This is one square unit.

What is the area of this square?

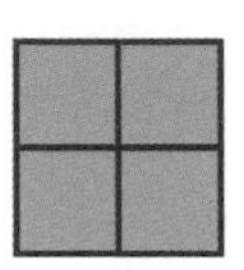

length × width = 2 × 2
4 square units

What is the area of this rectangle?

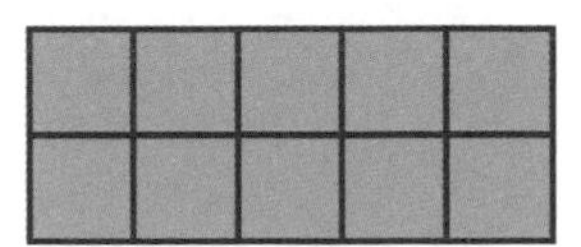

length × width = 5 × 2
10 square units

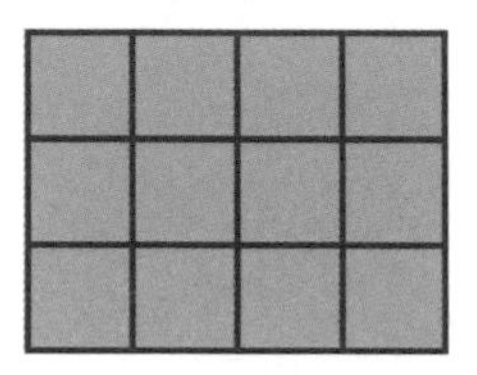

1. ________ square units

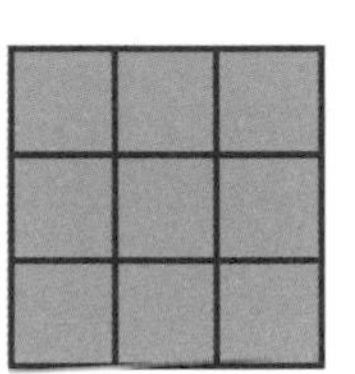

2. ________ square units

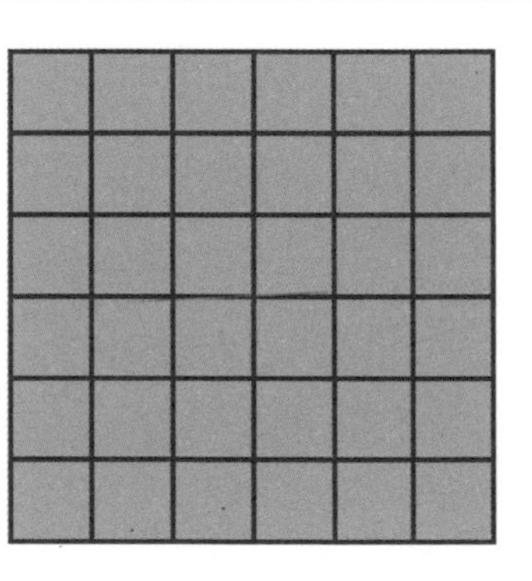

3. ________ square units

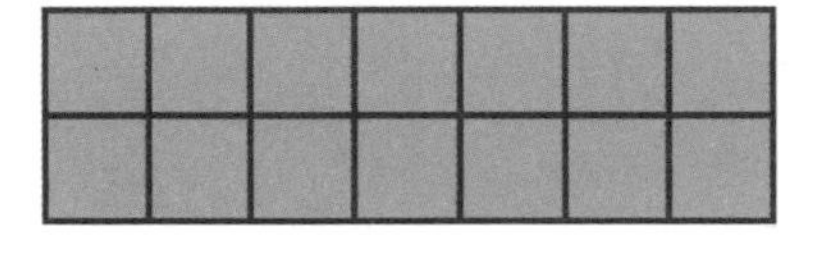

4. ________ square units

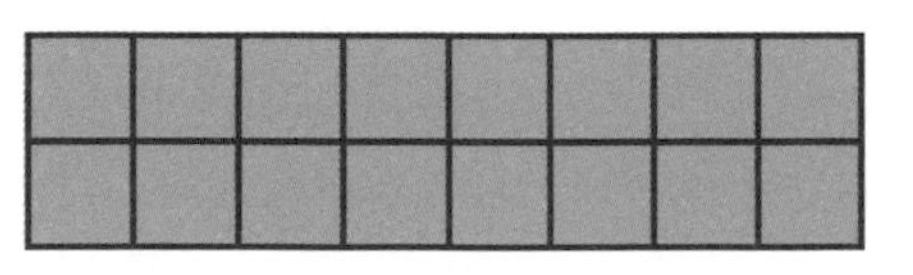

5. ________ square units

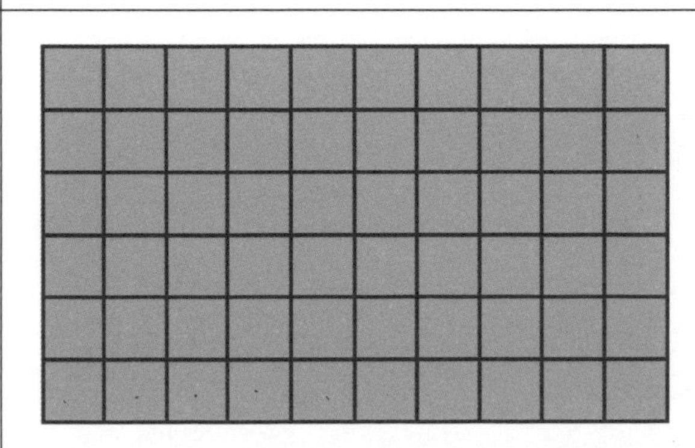

6. ________ square units

Squared Away

WRITE the area of each shape.

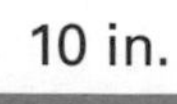

10 in.

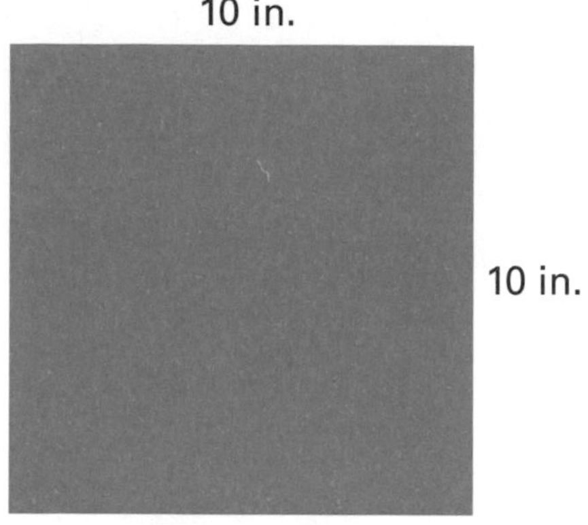

10 in.

1. _______ sq in.

7 in.

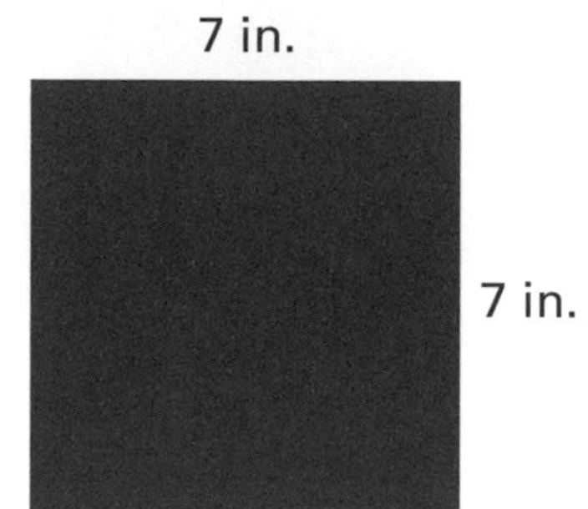

7 in.

2. _______ sq in.

11 ft

2 ft

3. _______ sq ft

3 ft

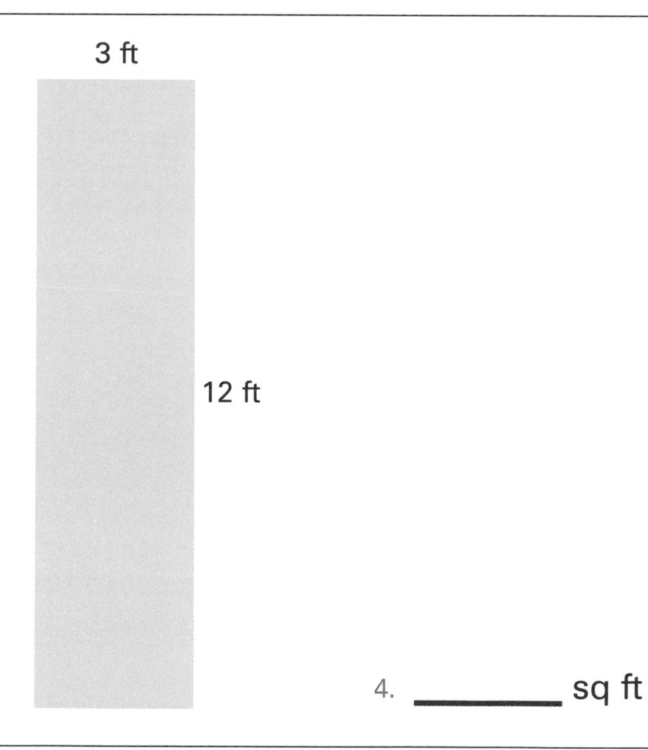

12 ft

4. _______ sq ft

9 cm

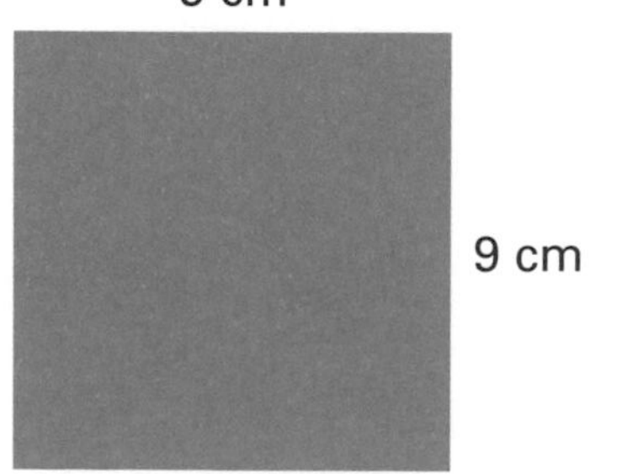

9 cm

5. _______ sq cm

15 in.

2 in.

6. _______ sq in.

Shape Creator

Each neighboring pair of dots is 1 centimeter apart. DRAW four different shapes that all have an area of 24 square centimeters.

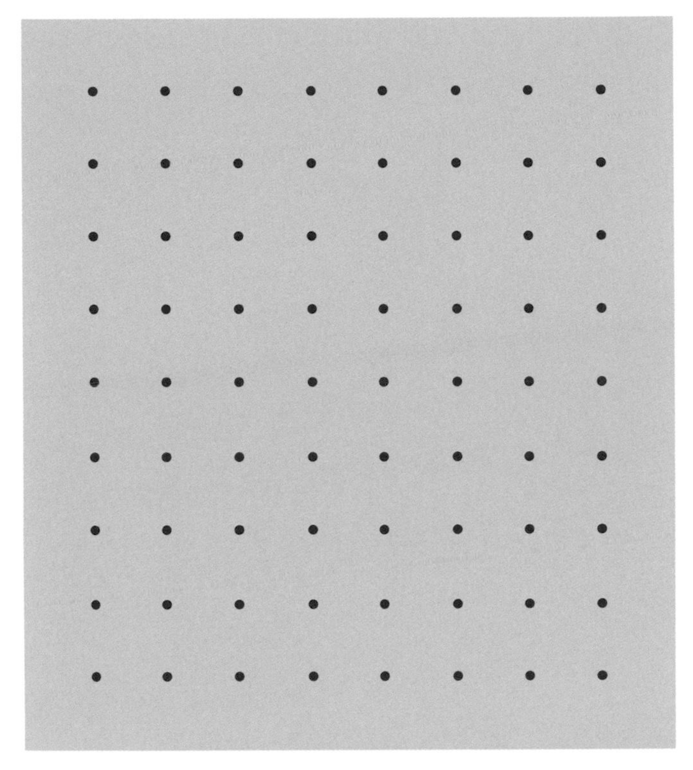

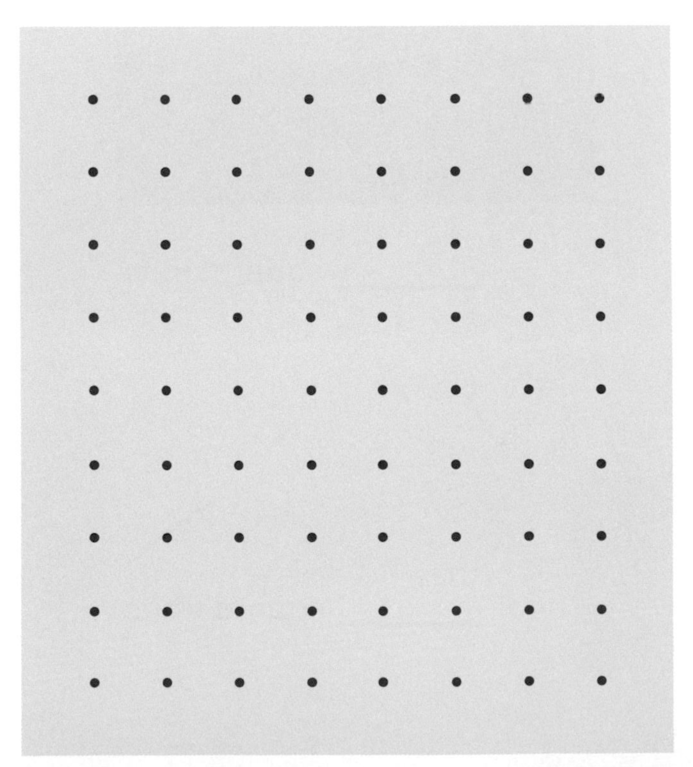

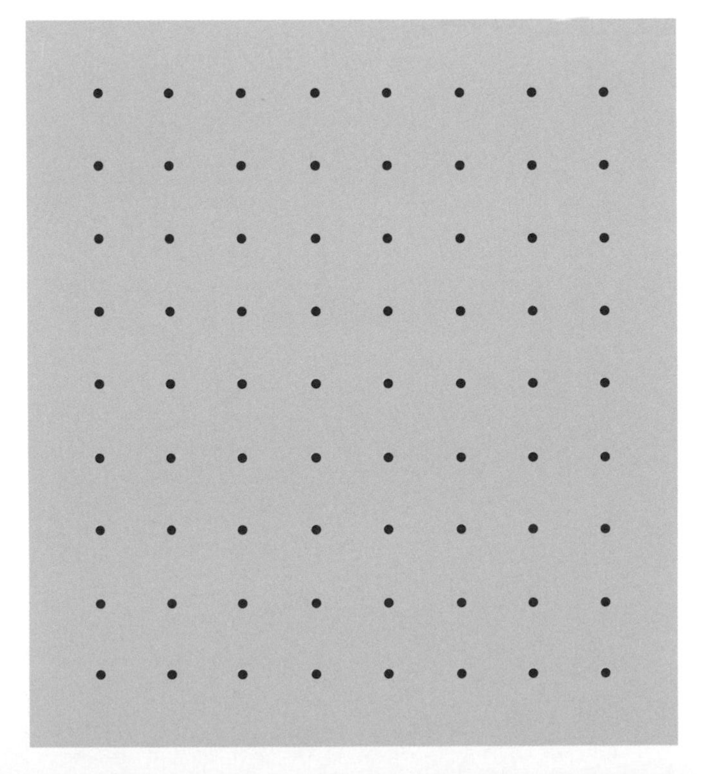

Puzzling Pentominoes

Using the pentomino pieces from page 115, PLACE the pieces to completely fill each shape without overlapping any pieces. Then WRITE the area of each shape. (Save the pentomino pieces to use again.)

HINT: Find the area as though the shape were complete, and then subtract the area of the missing pieces.

1. ________ square units

2. ________ square units

3. ________ square units

4. ________ square units

5. ________ square units

6. ________ square units

Missing Pieces

WRITE the area of each shape.

HINT: Find the area as though the shape were complete, and then subtract the area of the missing pieces.

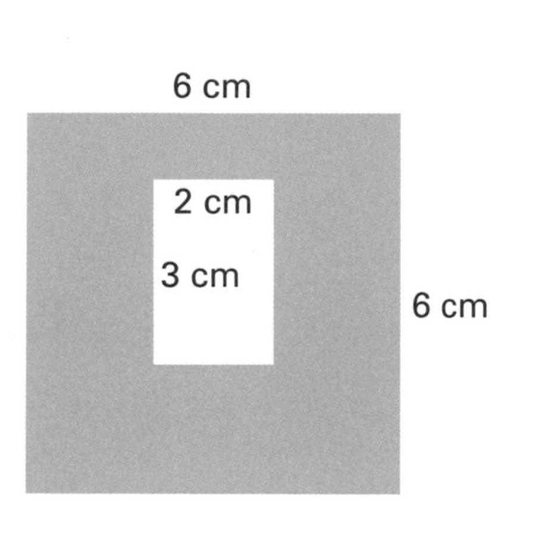

1. ________ sq cm

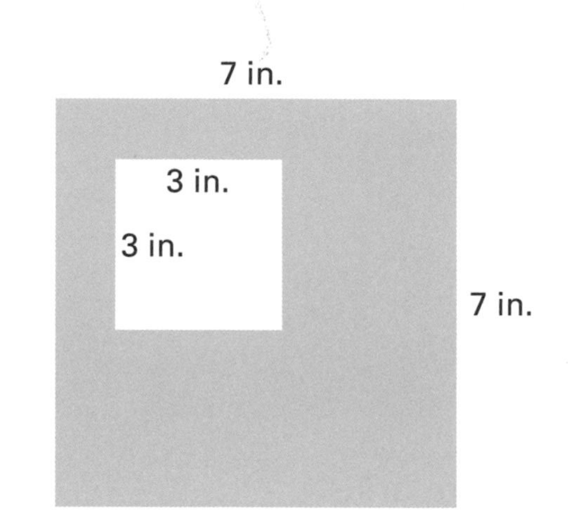

2. ________ sq in.

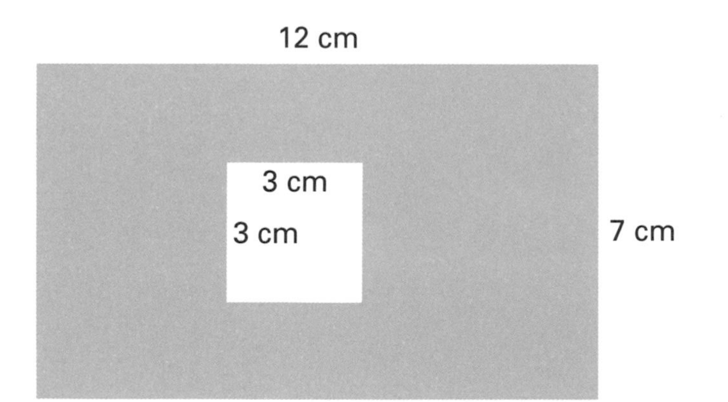

3. ________ sq cm

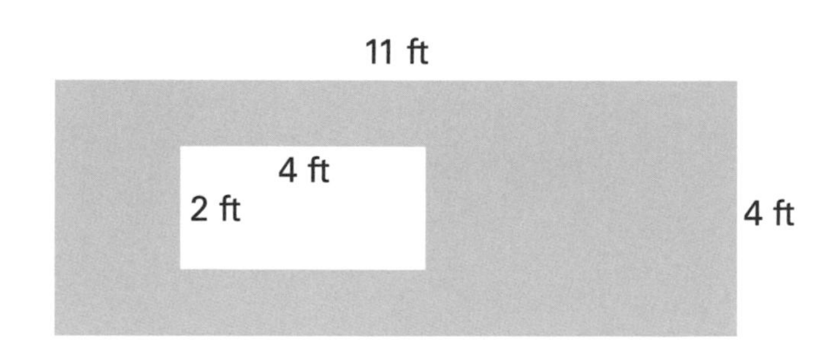

4. ________ sq ft

Unit Rewind

WRITE the perimeter of each shape.

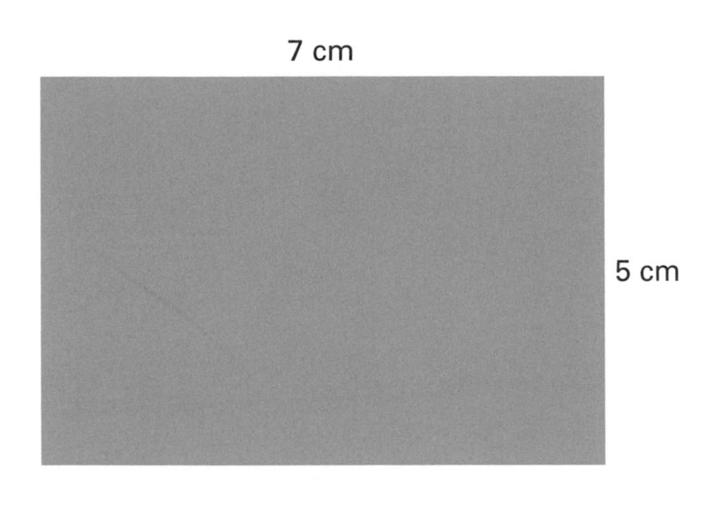

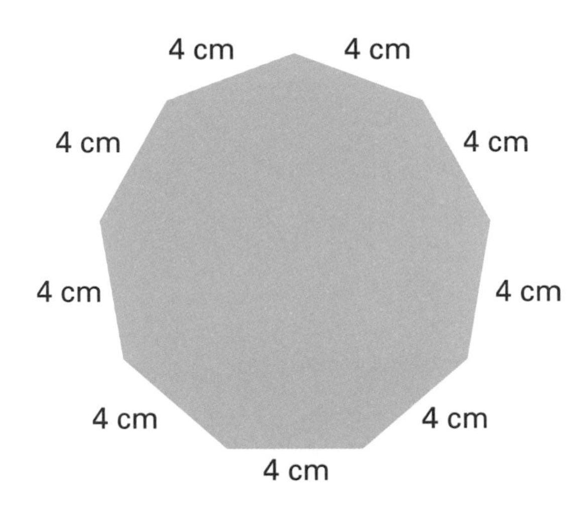

1. ________ cm

2. ________ cm

Each neighboring pair of dots is 1 centimeter apart. DRAW two different shapes that have a perimeter of 12 centimeters.

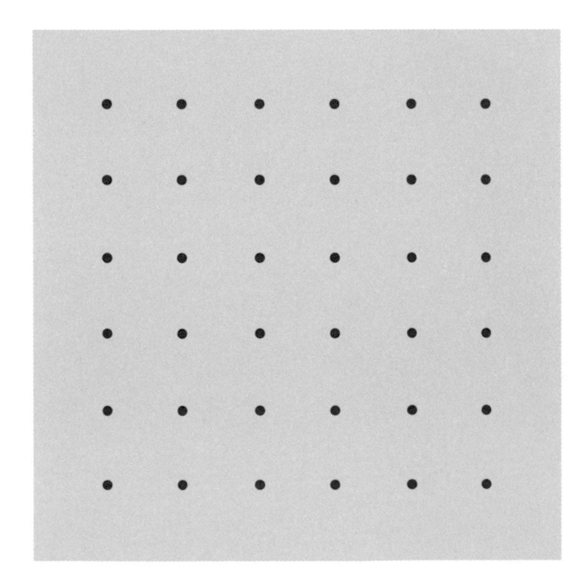

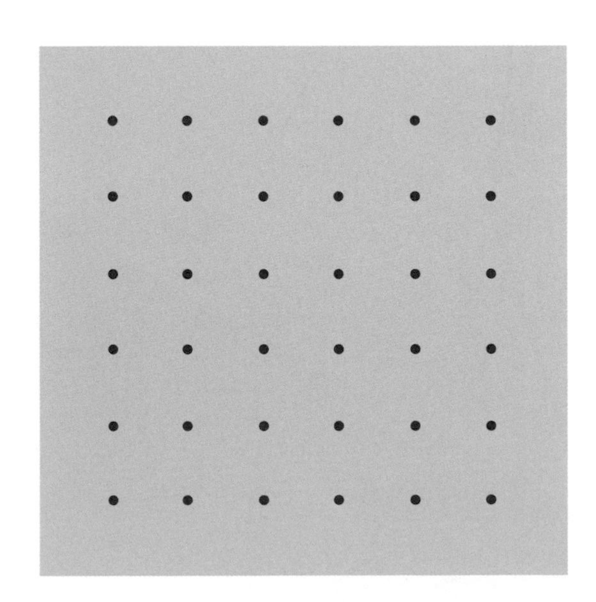

Unit Rewind

Using the measurements, WRITE the perimeter of each colored section.

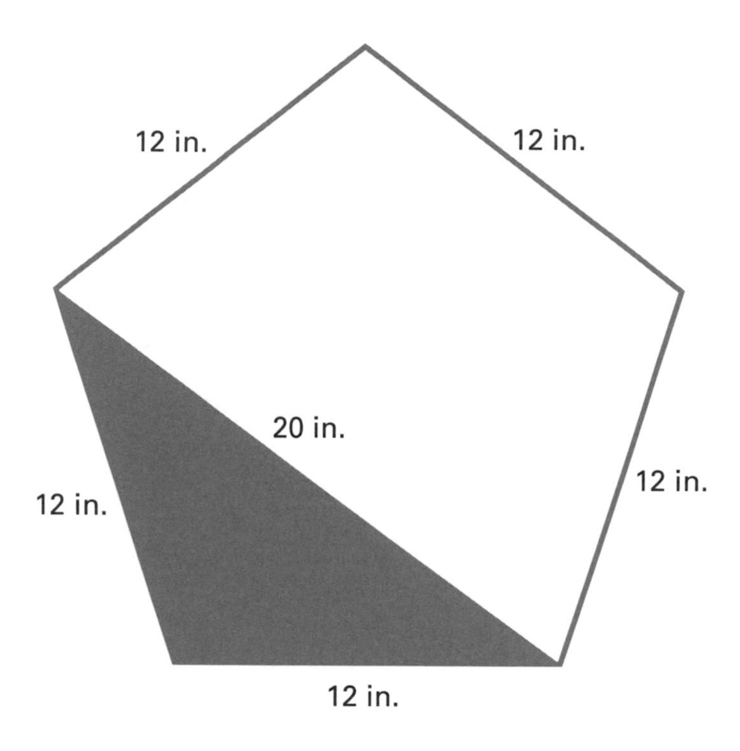

1. _______ in.

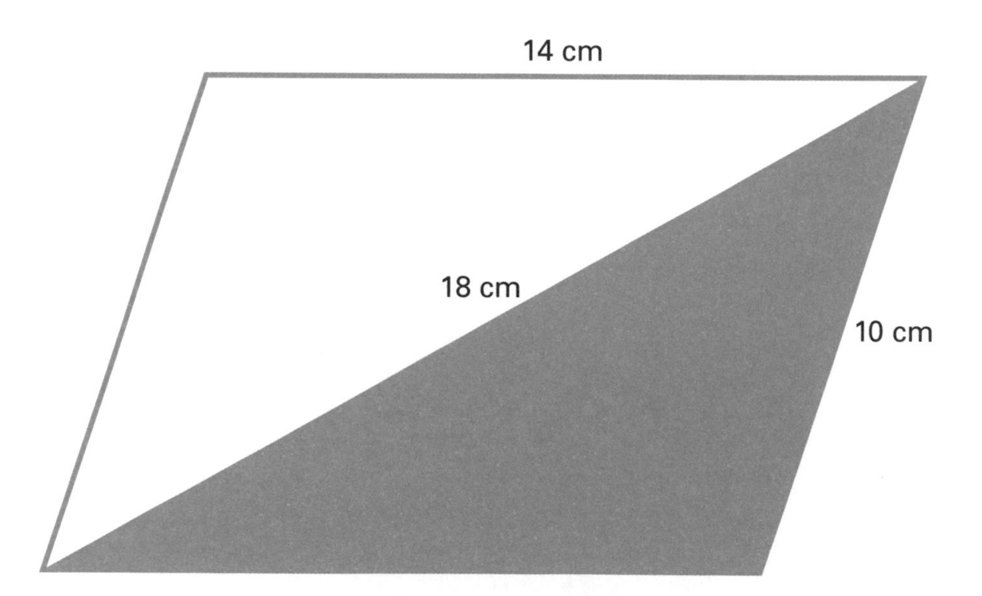

2. _______ cm

Unit Rewind

Using the pentomino pieces from page 115, PLACE the pieces to completely fill each shape without overlapping any pieces. Then WRITE the perimeter and area of each shape. (Save the pentomino pieces to use again.)

1. Perimeter: ______ units

   Area: ______ square units

2. Perimeter: ______ units

   Area: ______ square units

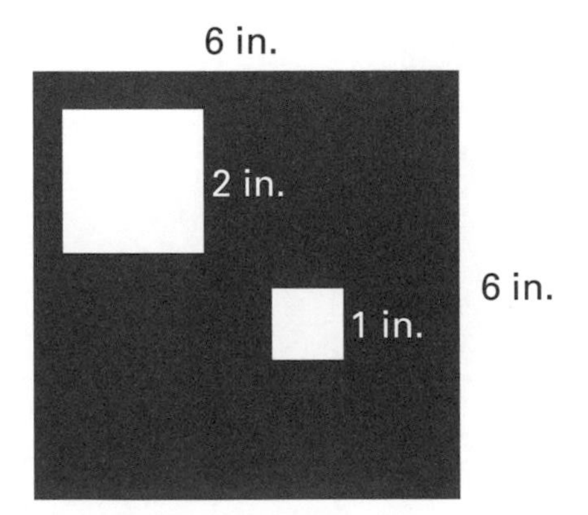

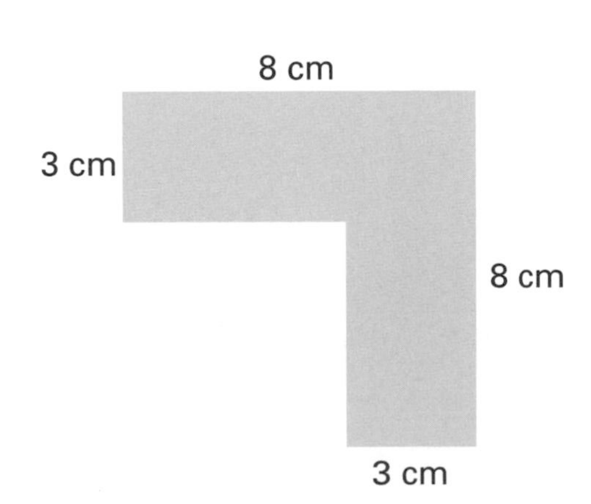

3. Perimeter: ______ in.

   Area: ______ square in.

4. Perimeter: ______ cm

   Area: ______ square cm

Puzzling Pentominoes

WRITE the perimeter and area of each shape.

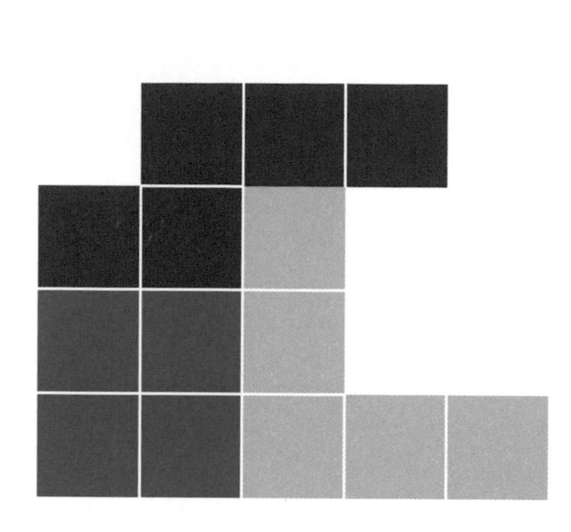

1. Perimeter: ________ units

   Area: ________ square units

2. Perimeter: ________ units

   Area: ________ square units

3. Perimeter: ________ units

   Area: ________ square units

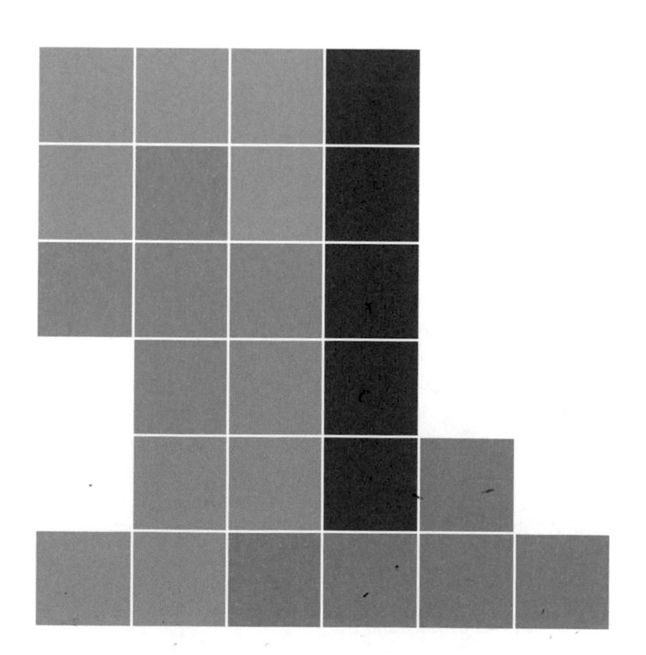

4. Perimeter: ________ units

   Area: ________ square units

Puzzling Pentominoes

Using the pentomino pieces from page 115, TRACE and COLOR two different ways to make a shape with a perimeter of 20 units. (Save the pentomino pieces to use again.)

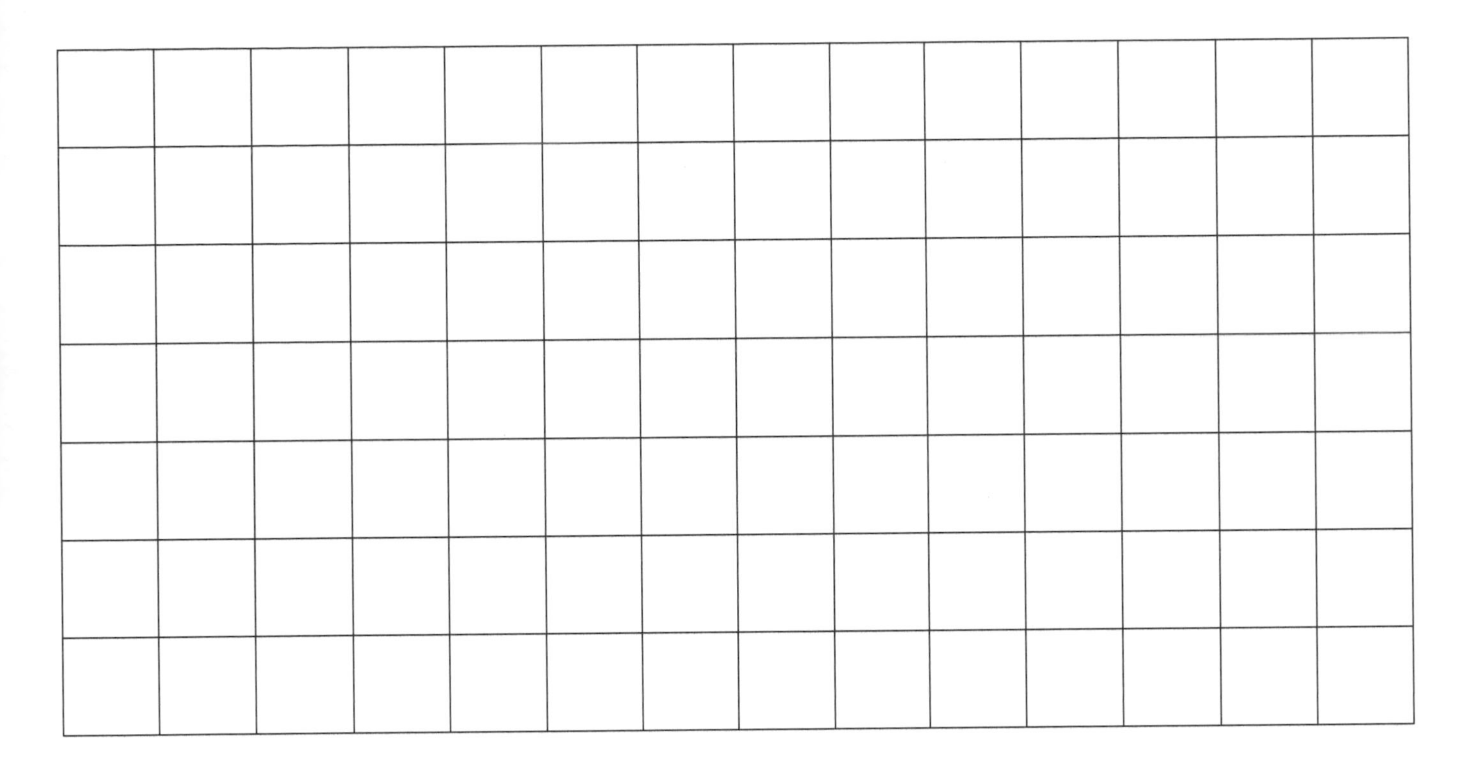

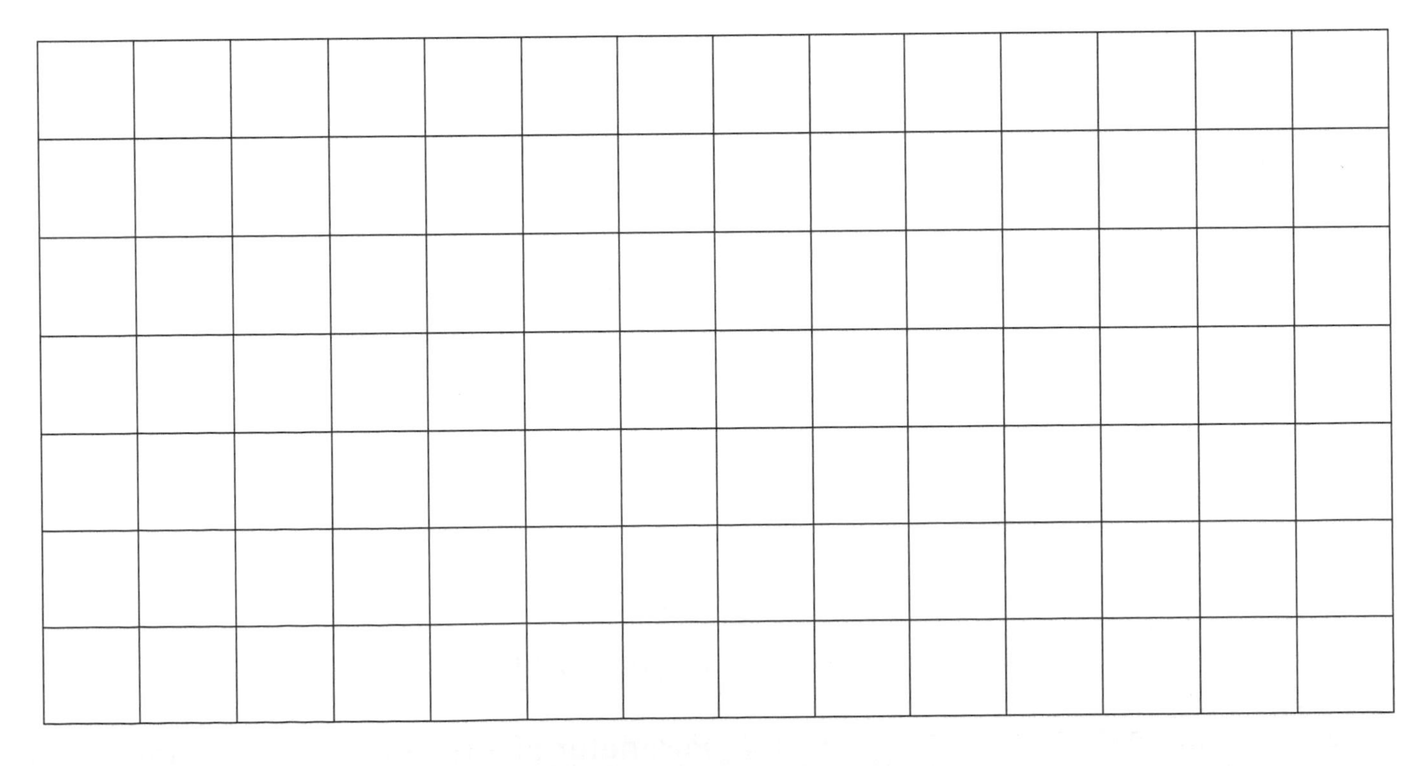

Puzzling Pentominoes

Using the pentomino pieces from page 115, TRACE and COLOR two different shapes. WRITE the perimeter of each shape. (Save the pentomino pieces to use again.)

Perimeter of shape 1: ________ units

Perimeter of shape 2: ________ units

Puzzling Pentominoes

Using the pentomino pieces from page 115, PLACE every piece on the board without overlapping any pieces. The board will not be completely filled. TRACE and COLOR your shape. WRITE the area and the perimeter of the shape. (Save the pentomino pieces to use again.)

Perimeter: ________ units

Area: ________ square units

Pentomino Place

Use the pentomino pieces from page 115. READ the rules. PLAY the game! (Save the pentomino pieces to use again.)

Rules: Two or three players

1. Take turns placing a pentomino piece on the board. Pieces cannot overlap.
2. Continue playing until no more pentomino pieces can fit on the board.

The player to place the last pentomino piece on the board wins!

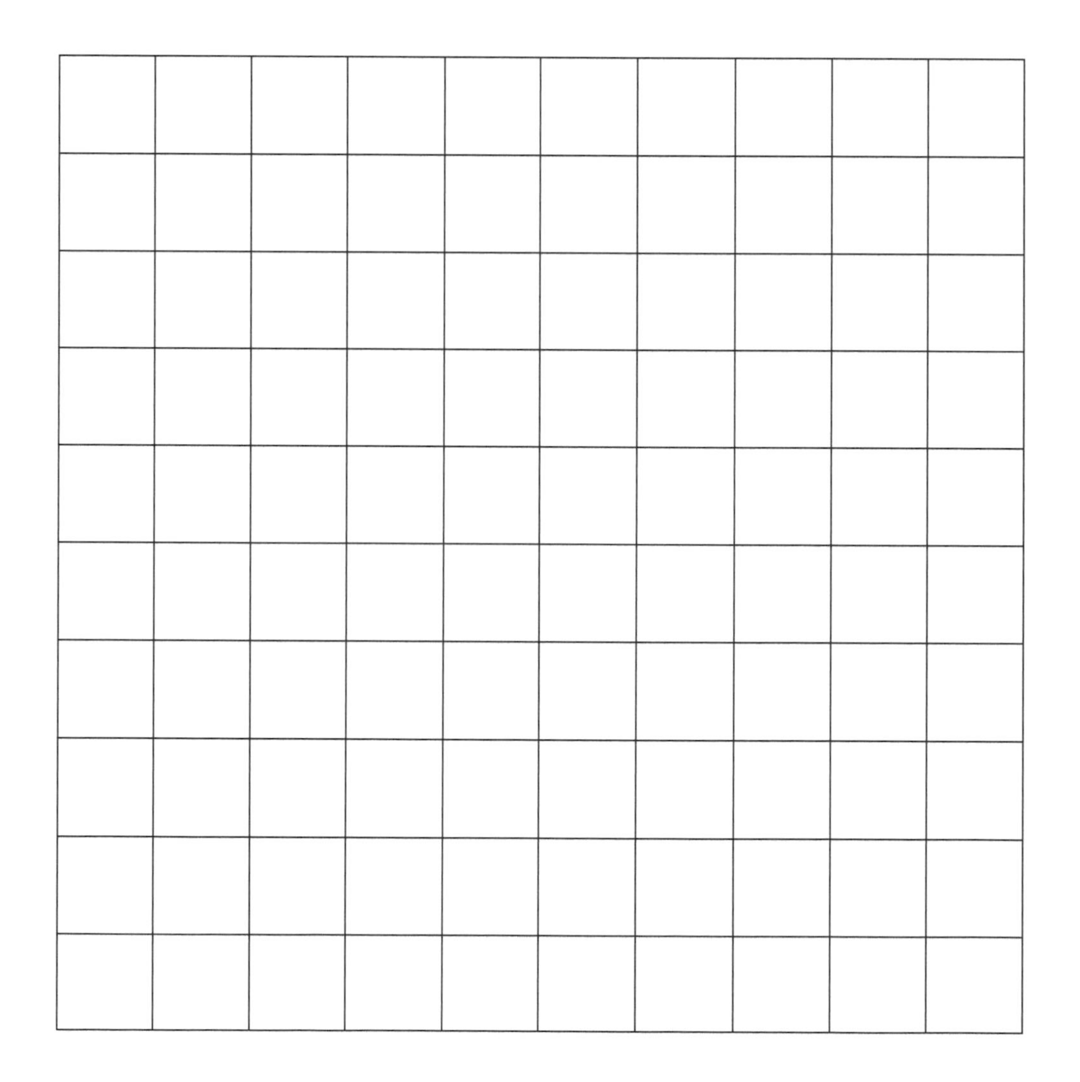

Tricky Tangrams

Using the tangram pieces from page 117, PLACE the pieces to completely fill each shape without overlapping any pieces. (Save the tangram pieces to use again.)

HINT: Try placing the biggest pieces first.

Tricky Tangrams

Using the tangram pieces from page 117, PLACE the pieces to completely fill each shape without overlapping any pieces. (Save the tangram pieces to use again.)

HINT: Try placing the biggest pieces first.

Tricky Tangrams

Using the tangram pieces from page 117, PLACE and DRAW two different shapes. Then challenge someone else to solve the puzzles you've made. (Save the tangram pieces to use again.)

Perfect Patterns

A **tessellation** is the pattern made when shapes repeat without gaps or overlapping. DRAW and COLOR the rest of each tessellation. Don't forget to extend in all directions!

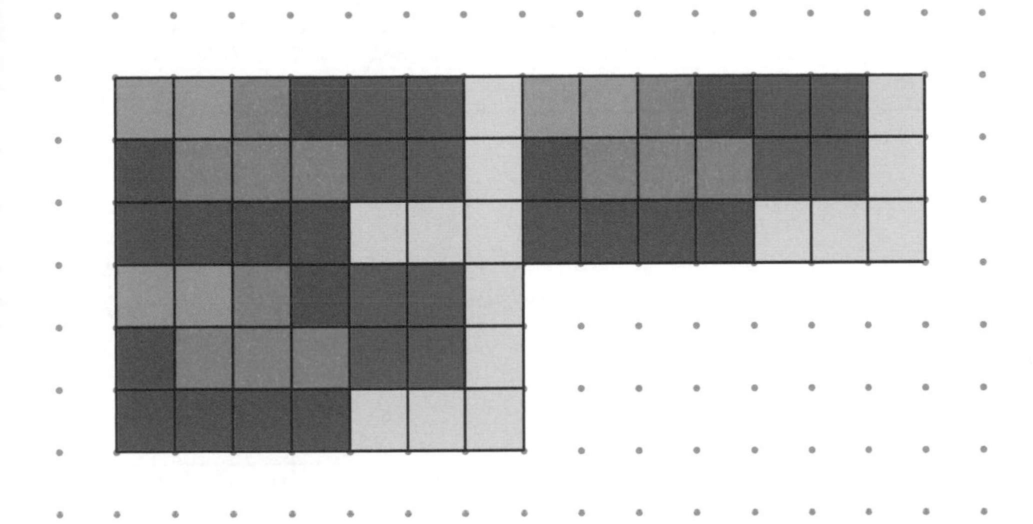

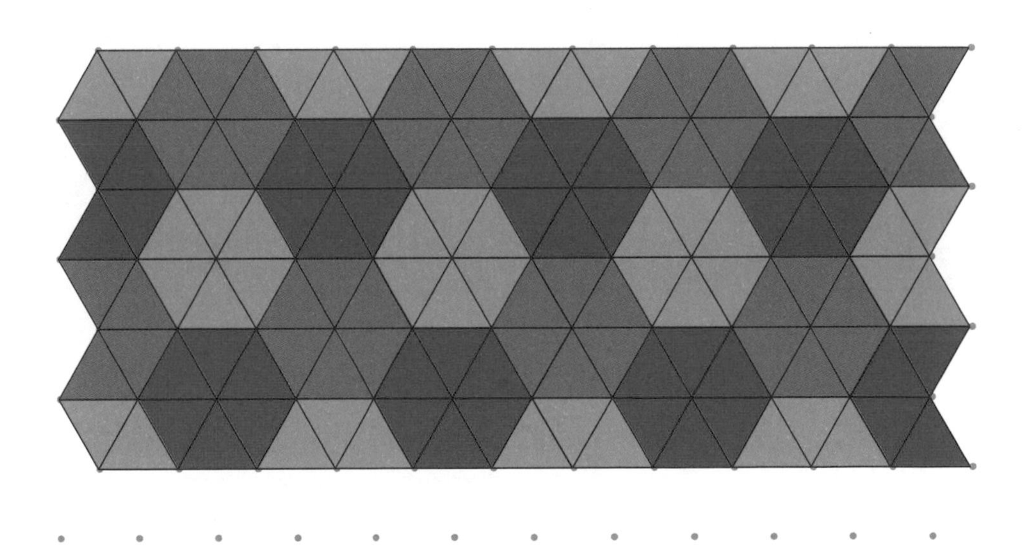

Perfect Patterns

DRAW and COLOR the rest of the tessellation.

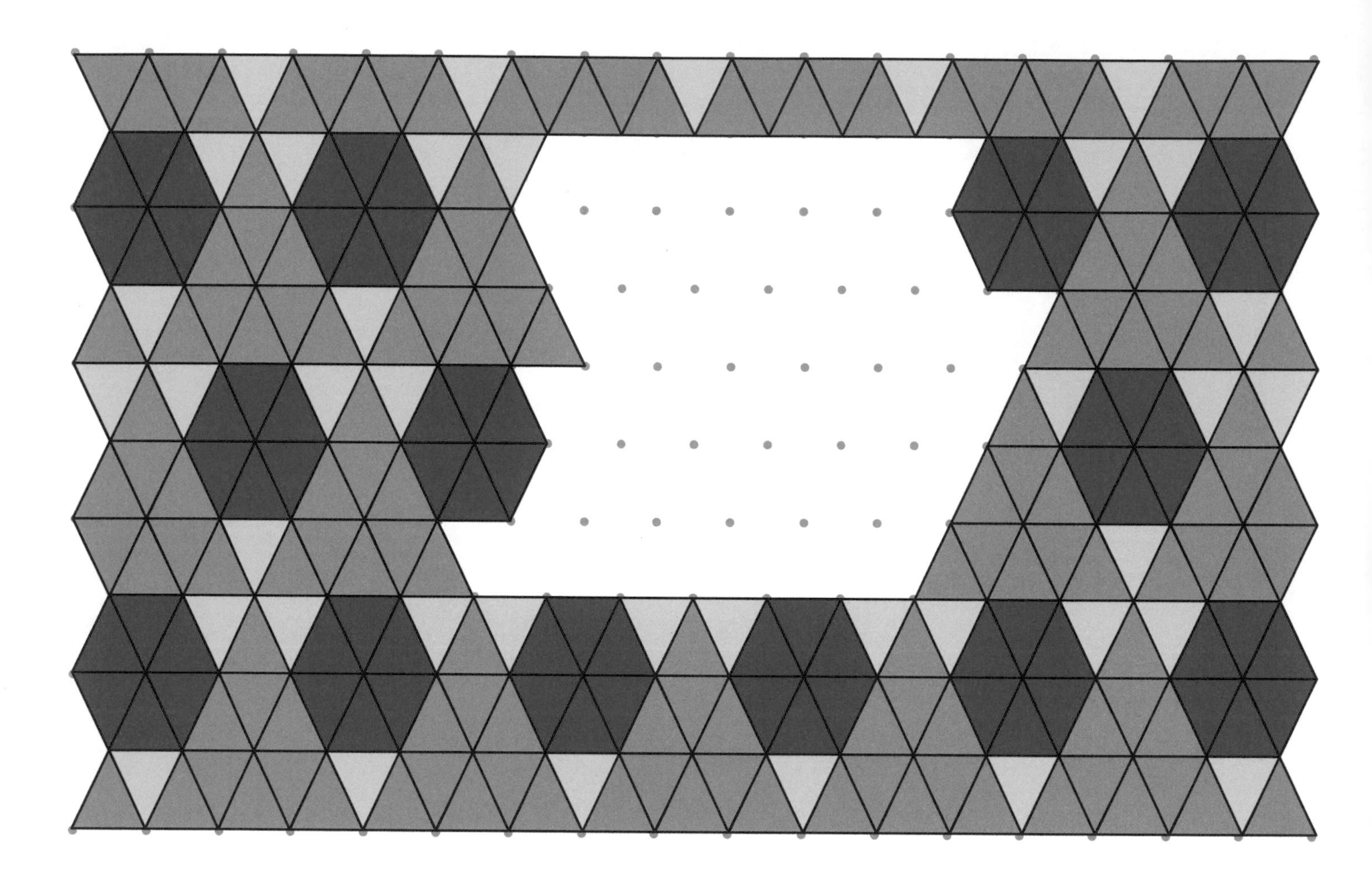

Perfect Pattern Play

READ the rules. MAKE a perfect pattern!

Rules: Two players
Player 1: Create and color a tessellation in the upper left part of the grid. Hand the tessellation to the other player.
Player 2: Extend the tessellation to the right until the upper right section of the grid is complete. Hand the tessellation back to the first player.
Player 1: Extend the tessellation until the lower left section of the grid is full. Hand the tessellation to the other player.
Player 2: Extend the tessellation until the rest of the grid is full.

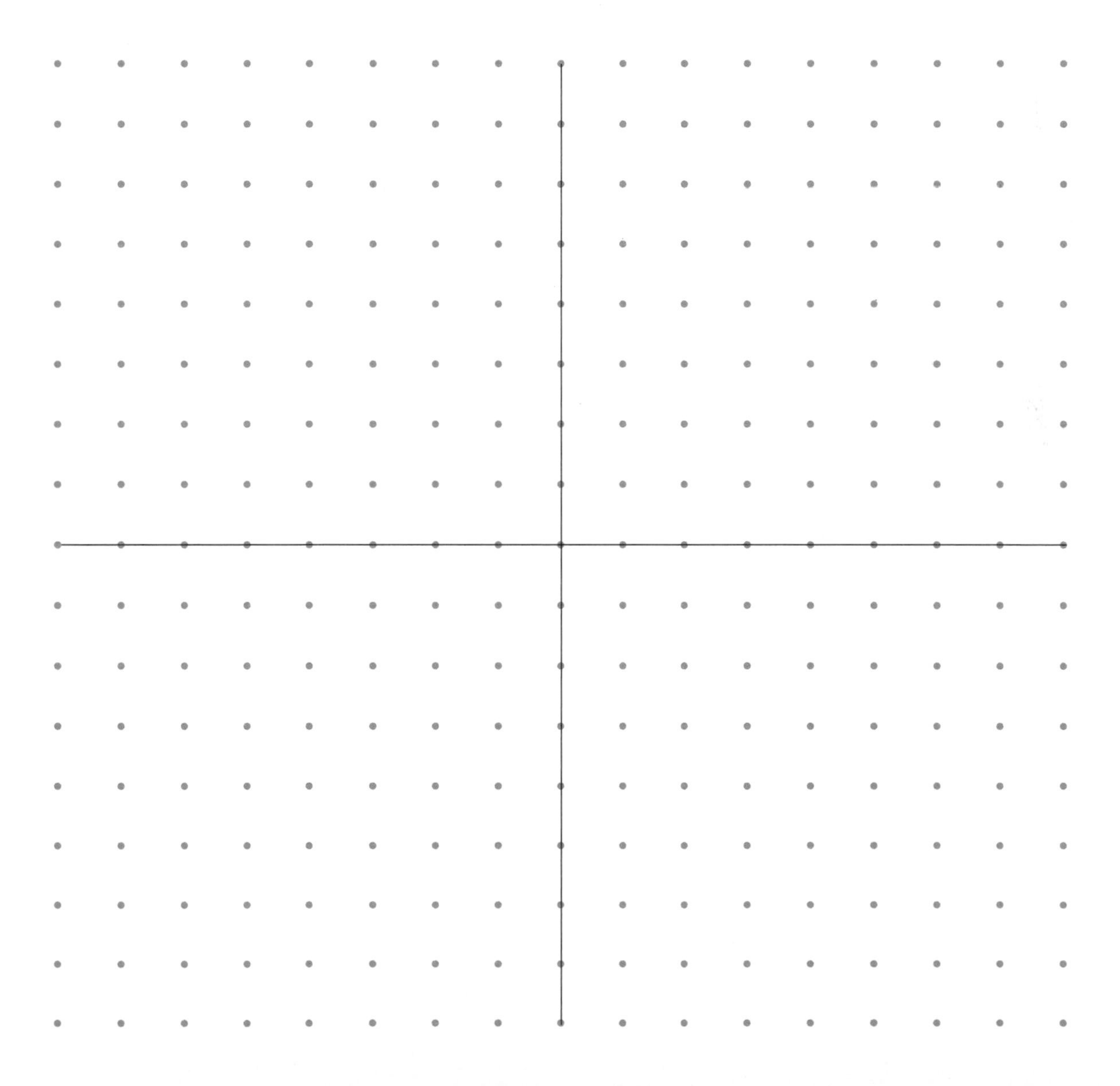

Tessellation Designer

Design your own tessellation! COLOR the triangles to make your own tessellation.

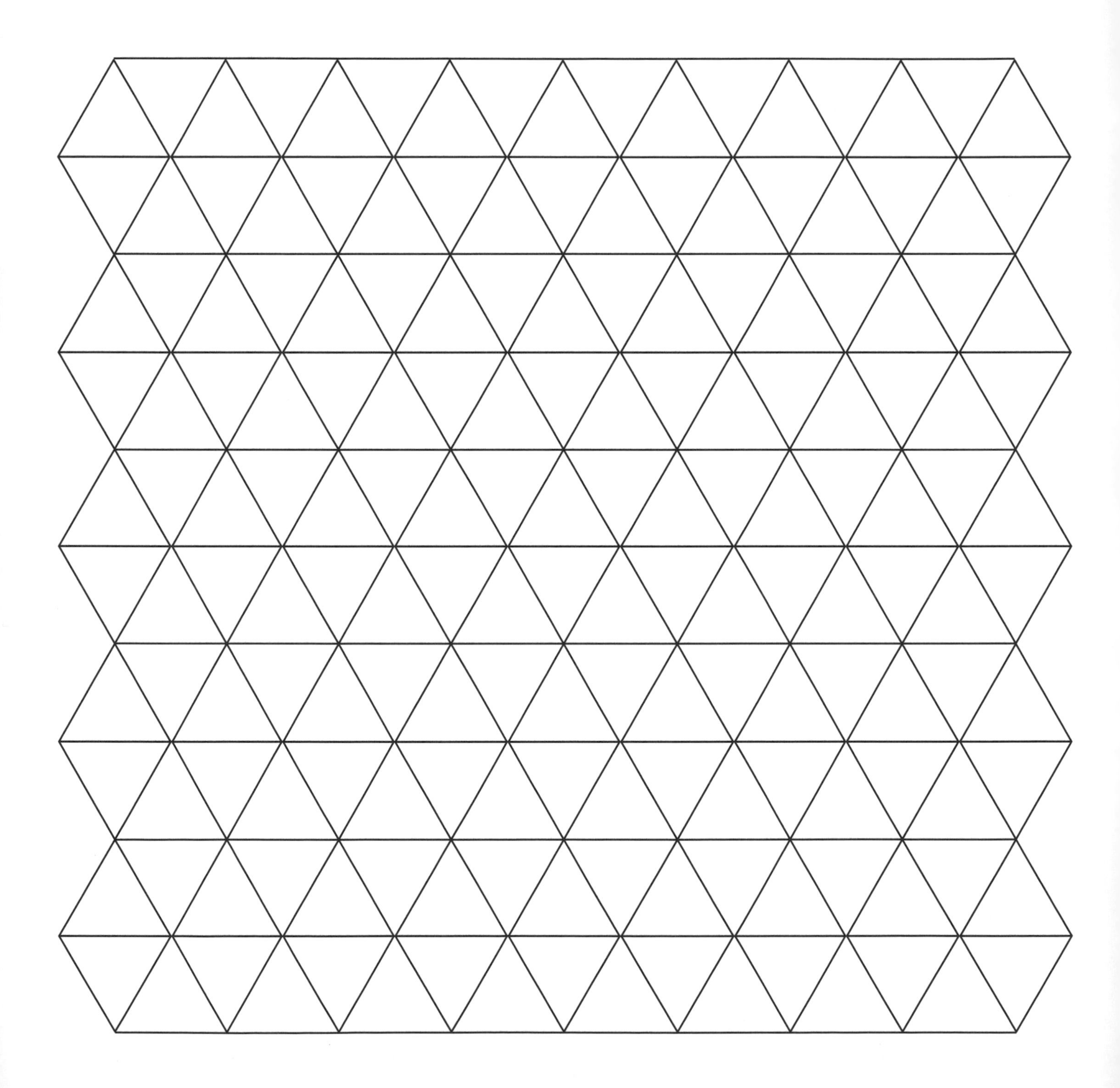

Puzzling Pentominoes

Using **all** of the pentomino pieces from page 115, TRACE and COLOR a shape with a perimeter of at least 40 units.

Perimeter: ________ units

Area: ________ square units

Puzzling Pentominoes

Using the pentomino pieces from page 115, TRACE and COLOR a shape with an area that is greater than 36 units. (Save the pentomino pieces to use again.)

Tricky Tangrams

Using the tangram pieces from page 117, PLACE the pieces to completely fill each shape without overlapping any pieces.

Tessellation Designer

DESIGN and COLOR your own tessellation.

Pentominoes

CUT OUT the 13 pentomino pieces.

These pentomino pieces are for use with pages 74, 75, 86, 87, 92, 93, 97, 99, 100, 101, 102, 111, and 112.

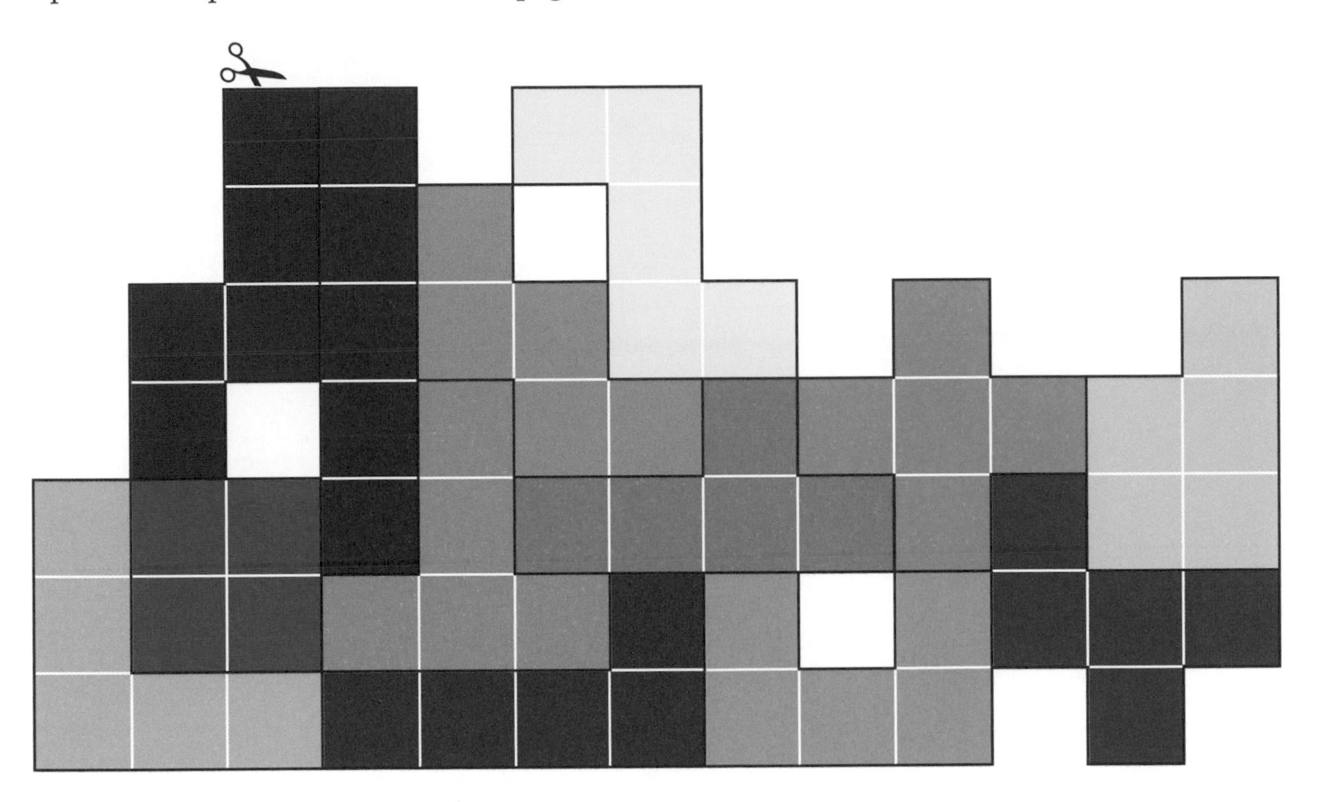

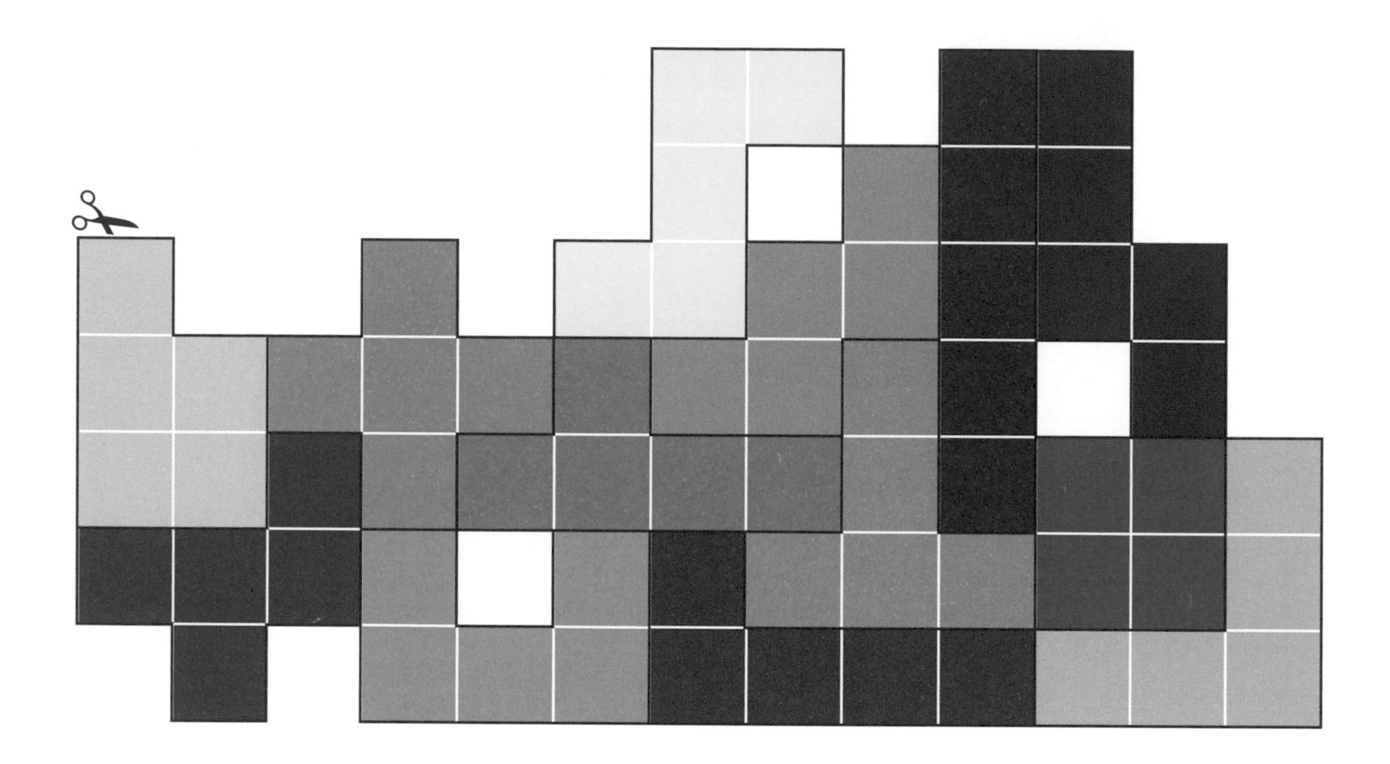

Tangrams

CUT OUT the seven tangram pieces.

These tangram pieces are for use with pages 103, 104, 105, 106, and 113.

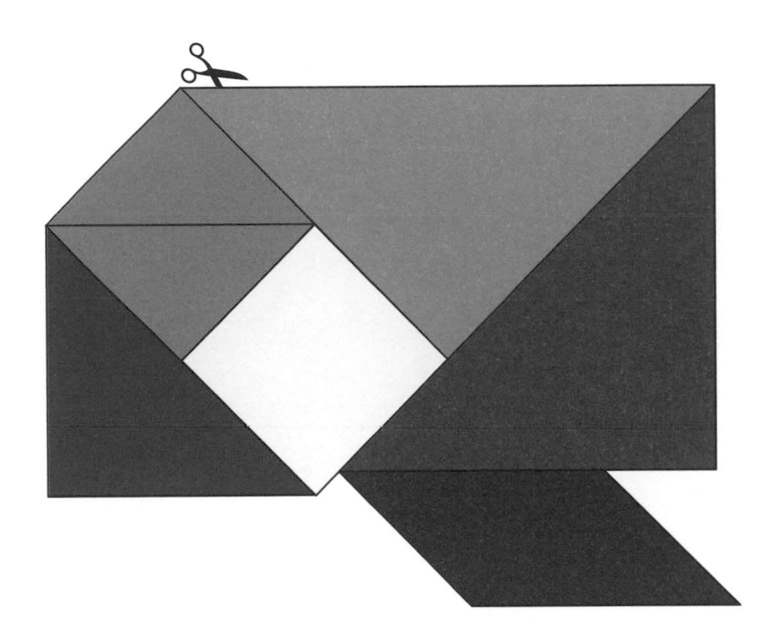

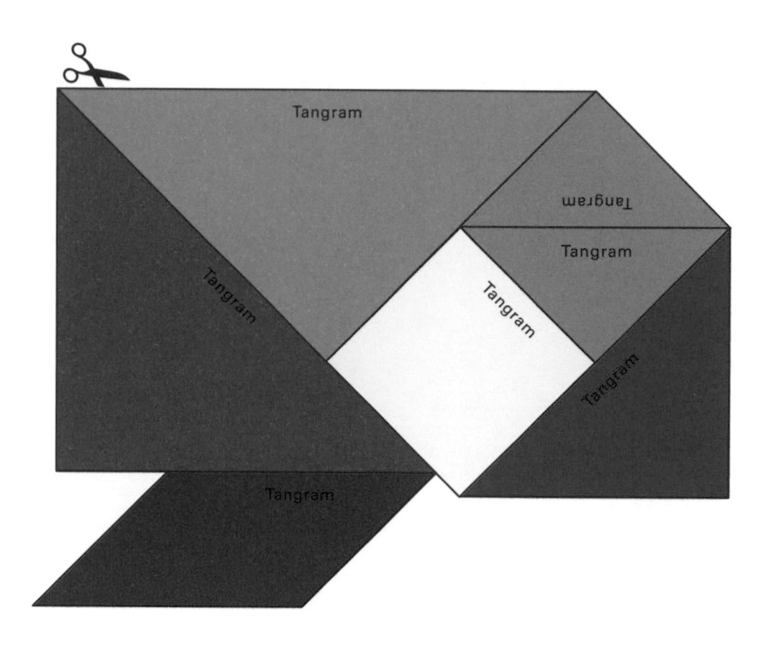
Tangram
Tangram
Tangram
Tangram
Tangram
Tangram
Tangram